# LET'S PREPARE FOR THE FCAT GRADE 8 MATH EXAM



## FLORIDA COMPREHENSIVE ASSESSMENT TEST

PAMELA WINDSPIRIT

National Board Certified
Math Program Planner
The Academy at Charlotte Technical Center
Port Charlotte, Florida



BARRON'S

## About the Author

**Pamela Windspirit** has taught mathematics on both the middle school and high school level in Charlotte County, Florida, for the last fifteen years. She received an M.Ed. in Middle Grades Curriculum and Instruction from the University of South Florida, and is National Board Certified in Early Adolescent Mathematics. Pamela has presented at numerous conferences throughout Florida and the United States. Currently, she teaches at Career Quest, a satellite of The Academy at Charlotte Technical Center, an alternative high school, and acts as the National Board for Professional Teaching Standards Coordinator for Charlotte County. She is also the author of Barron's *How to Prepare for the FCAT in High School Math*, which was published in 2003.



*All inquiries should be addressed to:*

Barron's Educational Series, Inc.
250 Wireless Boulevard
Hauppauge, New York 11788
**www.barronseduc.com**

ISBN-13: 978-0-7641-3193-6
ISBN-10: 0-7641-3193-1

Library of Congress Catalog Card No. 2004045318

**Library of Congress Cataloging-in-Publication Data**

Windspirit, Pamela J.
Let's prepare for the Florida Comprehensive Assessment Test (FCAT) grade 8 exam in mathematics / Pamela Windspirit.
p. cm.
Includes index.
ISBN 0-7641-3193-1
1. Mathematics—Examinations, questions, etc. 2. Florida Comprehensive Assessment Test—Study guides. 3. Mathematical ability—Testing. I. Title.

QA43.W573 2005
510'.76—dc22 2005045318

Printed in the United States of America

9 8 7 6 5 4 3 2 1

# Contents

# Introduction to the FCAT

## WHAT IS THE FCAT?

The Florida Comprehensive Assessment Test (FCAT) in Mathematics measures what students should know and be able to do by the eighth grade. It is based on a set of standards called the Sunshine State Standards that cover number sense, measurement, geometry, algebraic thinking, and data analysis and probability.

## HOW TO USE THIS BOOK

*Let's Prepare for the FCAT Grade 8 Exam in Mathematics* will help you do your best on the mathematics portion of the FCAT. This book reviews what you have learned in math and helps you become familiar with the types of questions on the FCAT and the way these questions are asked.

Each lesson deals with a particular topic covered on the FCAT and gives some sample problems similar to those that may appear on the FCAT. At the end of the book are two practice tests with explained answers.

## TYPE OF QUESTIONS

There are four types of questions on the FCAT: multiple choice, gridded response, short response, and extended response.

### Multiple Choice

- Are worth 1 point each.
- Should take about 1 minute to solve each problem.
- Give four answer choices (A, B, C, D or F, G, H, I).

### Gridded Response (at right)

- Are worth 1 point each.
- Should take about 1.5 minutes to solve each problem.
- Require you to bubble in the answer. Answers should be bubbled all the way to the left or all the way to the right with no spaces between.

- Answers are never negative.
- Use only one blank box per number or symbol, and do not add boxes or bubbles.
- A decimal point takes up an entire box and should be filled in.

## Short Response

- Have a Think-Solve-Explain box next to the problem.
- Are worth 2 points each.
- Should take about 5 minutes to solve each problem.
- Usually have two parts, Part A and Part B, that require you to write an answer, explain an answer, or show your work.
- Allow you to receive partial credit if you are able to do only part of the problem correctly.

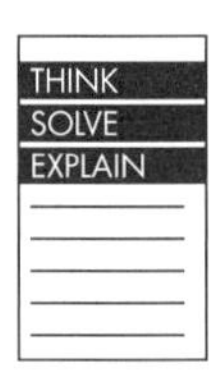


## Extended Response

- Have a longer Think-Solve-Explain box next to the problem.
- Are worth 4 points each.
- Should take about 10 minutes to solve each problem.
- Usually have three parts, Part A, Part B, and Part C, that require you to write an answer, explain an answer, or show your work.
- Allow you to receive partial credit (1, 2, or 3 points) if you are able to do only part of the problem correctly.

# FCAT REFERENCE SHEETS

During the testing period you will be supplied with conversions and formulas that you may need to solve measurement problems.

## Conversions for the FCAT

| Metric | US Customary |
|---|---|
| 1 kilometer = 1000 meters<br>1 meter = 100 centimeters = 1000 millimeters<br>1 liter = 1000 milliliters = 1000 cubic centimeters<br>1 gram = 1000 milligrams<br>1 kilogram = 1000 grams<br><br>**Time**<br>1 hour = 60 minutes<br>1 minute = 60 seconds | 1 mile = 1760 yards = 5280 feet<br>1 yard = 3 feet = 36 inches<br>1 acre = 43,560 square feet<br>1 gallon = 4 quarts = 8 pints<br>1 quart = 2 pints = 4 cups<br>1 pint = 2 cups = 16 ounces<br>1 cup = 8 fluid ounces<br>1 ton = 2000 pounds<br>1 pound = 16 ounces |

## Formulas for the FCAT

| Figure | Formula |
|---|---|
| Triangle | Area = $\frac{1}{2}bh$ |
| Rectangle | Area = $lw$ |
| Trapezoid | Area = $\frac{1}{2}(b_1 + b_2)h$ |
| Parallelogram | Area = $bh$ |
| Cirle | Area = $\pi r^2$<br>$C = \pi d = 2\pi r$ |
| Right circular cylinder | Volume = $\pi r^2 h$ or volume = $Bh$<br>Surface area = $2\pi rh + 2\pi r^2$ |
| Rectangular solid | Volume = $lwh$ or volume = $Bh$<br>Surface area = $2(lw) + 2(hw) + 2(lh)$ |

$b$ = base; $h$ = height; $l$ = length; $w$ = width; $r$ = radius; $d$ = diameter; C = circumference;

$\ell$ = slant height; $\pi = 3.14$ or $\frac{22}{7}$; $B$ = area of base of figure

# Lesson 1

## Whole Numbers and Decimals

What is the value of the 7 in the number 30,178,001?

The 7 is in the tens place in the thousands period; therefore the value of the 7 is 7 times 10,000, or 70,000.

Place Value for Whole Numbers

| Billions Period | | | Millions Period | | | Thousands Period | | | Ones Period | | |
|---|---|---|---|---|---|---|---|---|---|---|---|
| Hundreds | Tens | Ones | Hundreds | Tens | Ones | Hundreds | Tens | Ones | Hundreds | Tens | Ones |
| $10^{11}$ | $10^{10}$ | $10^9$ | $10^8$ | $10^7$ | $10^6$ | $10^5$ | $10^4$ | $10^3$ | $10^2$ | $10^1$ | $10^0$ |
| | | | | 3 | 0 | 1 | 7 | 8 | 0 | 0 | 1 |

## PLACE VALUES FOR INTEGERS AND DECIMALS

### Integers

*Place value* tells you the *value* of a digit within a number. Each place in our number system has a value of 10 times the number to its right. In the number 34,567, the 4 is located in the thousands position, and so it has a value of 4 times 1000, or 4000. The number 5 has a value of 5 times 100, or 500.

Place value works with negative numbers too. For example, in the number –532, the 3 has a value of –30 while the 5 has a value of –500. Negative 30 is larger than negative 500 because it is further to the right on the number line.

## Decimal Fractions

Decimal fractions follow a pattern very similar to that for whole numbers. On the place-value chart you see that 10 is represented by a positive power of 10 while one tenth is represented by a negative power of 10. For example, one tenth = 0.1 or $\frac{1}{10}$. One tenth can also be written in exponential form as $10^{-1}$. Notice the pattern: one hundred = $10^2$, while one hundredth can be written as $10^{-2}$, and one thousandth = 0.001 or $\frac{1}{1000}$, and can also be written as $10^{-3}$.

**Place Value Chart**

| Exponential Form | Decimal Fraction | Standard Form | Word Form |
|---|---|---|---|
| $10^6$ | | 1,000,000 | Millions |
| $10^5$ | | 100,000 | Hundred thousands |
| $10^4$ | | 10,000 | Ten thousands |
| $10^3$ | | 1,000 | Thousands |
| $10^2$ | | 100 | Hundreds |
| $10^1$ | | 10 | Tens |
| $10^0$ | | 1 | Ones |
| $10^{-1}$ | $\frac{1}{10}$ | 0.1 | Tenths |
| $10^{-2}$ | $\frac{1}{100}$ | 0.01 | Hundredths |
| $10^{-3}$ | $\frac{1}{1000}$ | 0.001 | Thousandths |
| $10^{-4}$ | $\frac{1}{10,000}$ | 0.0001 | Ten thousandths |
| $10^{-5}$ | $\frac{1}{100,000}$ | 0.00001 | Hundred thousandths |

**Important! A negative exponent does not mean that the number it represents is negative!**

## READING AND WRITING WHOLE NUMBERS AND DECIMALS

Numbers can be written in *standard form*, which is the way you normally see them. They can also be written in word form, expanded form, or exponential form. For expanded form, use the place value to write the number. For exponential form, multiply the digit by the power of 10 for the place value.

| Form | Sample |
|---|---|
| Standard form | 45,617 |
| Word form | Forty-five thousand six hundred seventeen |
| Expanded form | $(4 \times 10{,}000) + (5 \times 1000) + (6 \times 100) + (1 \times 10) + (7 \times 1)$ |
| Exponential form | $(4 \times 10^4) + (5 \times 10^3) + (6 \times 10^2) + (1 \times 10^1) + (7 \times 10^0)$ |

| Form | Sample |
|---|---|
| Standard form | 100.02 |
| Word form | One hundred and two hundredths |
| Expanded form | $(1 \times 100) + (2 \times 0.01)$ |
| Exponential form | $(1 \times 10^2) + (2 \times 10^{-2})$ |

**Example 1:** Write the number 25,134 in expanded notation.

Numbers can be written in expanded notation without exponents and with exponents.

$25{,}134 = 20{,}000 + 5000 + 100 + 30 + 4$

Expanded Form Without Exponents

$(2 \times 10{,}000) + (5 \times 1000) + (1 \times 100) + (3 \times 10) + (4 \times 1)$

Exponential Form

$(2 \times 10^4) + (5 \times 10^3) + (1 \times 10^2) + (3 \times 10^1) + (4 \times 10^0)$

**Example 2:** Write the number five million four hundred thousand fifteen in standard notation.

Start with: $5 \times 1 \text{ million} + 4 \times 100 \text{ thousand} + 15 \times 1$

$= (5 \times 1{,}000{,}000) + (4 \times 100{,}000) + (15 \times 1)$

$= 5{,}000{,}000 + 400{,}000 + 15$

$= 5{,}400{,}015$

**Example 3:** Write the number 83,200 in exponential form.

$= (8 \times 10^4) + (3 \times 10^3) + (2 \times 10^2)$

**Example 4:** Write the number 0.25 in words.

= twenty-five hundred<u>ths</u>

**Example 5:** Write the number 500.12 in words.

= five hundred *and* twelve hundredths (Use *and* to represent a decimal point.)

**Example 6:** Write 24 ten thousandths as a decimal and as a fraction.

= 0.0024 and $\frac{24}{10,000}$

## ROUNDING

Round 275 to the nearest *hundred*.

The number 275 is between 200 and 300, but it is closer to 300 than to 200. Therefore, 275 rounds to 300.

Rounding numbers is a very important skill on the FCAT. During the test you may be asked to calculate an answer and then round it. If you can do the calculation but are weak in rounding, you may miss points.

You can round numbers by using a number line.

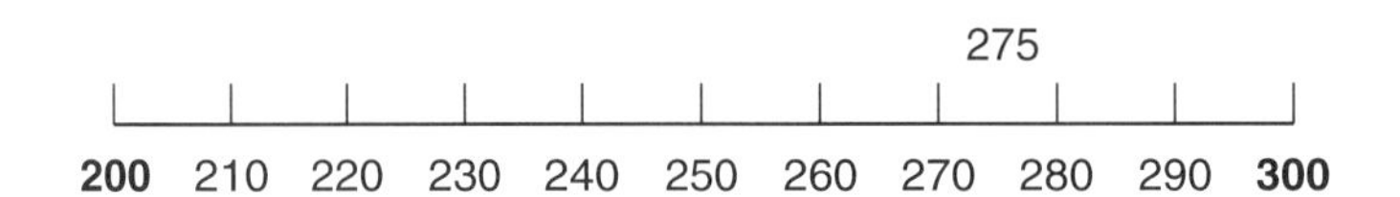


Another method of rounding whole numbers is to use place value.

Step 1: Underline the place value to which you are rounding.

Step 2: If the number to the immediate right of the underlined number is 5 or greater, round your underlined place up by one. All numbers to the right will now become zeros. If the number to the immediate right of the underlined number is less than 5, leave the underlined number the same and change all numbers to its immediate right to zeros.

**Example 1:** Round 143,459 to the nearest thousand.

143,459 is rounded to 143,000.

**Example 2:** Round 89,561 to the nearest hundred.

89,561 is rounded to 89,600.

**Example 3:** Round 799 to the nearest hundred.

799 is rounded to 800.

**Example 4:** Round 0.7534 to the nearest thousandth.

0.75<u>3</u>4 is rounded to 0.7530 = 0.753 (Drop trailing zeros for decimal rounding).

**Example 5:** Round 0.9851 to the nearest hundredth.

0.9<u>8</u>51 is rounded to 0.99.

**Example 6:** Round 0.599 to the nearest hundredth.

0.5<u>9</u>9 is rounded to 0.60.

## POWERS AND EXPONENTS

Powers are a way of writing repeated multiplication. For example, $5^3$, read as *five to the third power*, means to multiply 5 three times: $5 \times 5 \times 5$. The number 5 is called the *base*, and the 3 is the *exponent*.

**Example 1:** What is the value of $2^4 + 3^2$?

$2^4 = 2 \times 2 \times 2 \times 2 = 16$
$3^2 = 3 \times 3 = 9$
$16 + 9 = 25$

**Example 2:** Simplify $(a^3b^4)^2$.

$(a^3b^4) \cdot (a^3b^4) = a^3 \cdot a^3 \cdot b^4 \cdot b^4 = a^6b^8$

## SCIENTIFIC NOTATION FOR WHOLE NUMBERS

*Scientific notation* is a way of writing a number as the *product* of a power of 10 and a decimal number. The decimal number must be equal to or greater than 1 but less than 10. Use positive powers of 10 for numbers that are equal to or larger than 1. Use negative powers of 10 for numbers that are less than 1.

**Example 1:** Write 43,000 in scientific notation.

Step 1: Place a decimal point after the first digit (4): 4.3000.

Step 2: Multiply by 10 raised to the fourth power. $4.3 \times 10^4$. The decimal moved four places from its original position after the last zero to its new position after the 4; therefore use $10^4$.

**Example 2:** Write 0.00043 in scientific notation.

Step 1: Place a decimal point after the first nonzero digit (4), dropping the leading zeros: 4.3.

Step 2: Multiply by 10 raised to the negative fourth power: $4.3 \times 10^{-4}$. The decimal point has moved four places from its original position after the last zero to its new position after the 4; therefore use $10^{-4}$.

**Example 3:** Light travels about 186,000 miles per second. Write 186,000 in scientific notation.

$1.86 \times 10^{5}$. Move the decimal point so that it is between 1 and 8. The decimal point has moved five spaces, so multiply by 10 to the fifth power.

**Example 4:** An ant measures 0.078 inch in length. Write 0.078 in scientific notation.

$7.8 \times 10^{-2}$. Move the decimal point so that it is between 7 and 8. The decimal point has moved two spaces, so multiply by 10 to the negative second power.

**Example 5:** The radius of Earth is approximately $6.38 \times 10^{6}$ meters. Write the radius in standard notation.

Since the power of 10 is 6, move the decimal point in 6.38 six places to the right, adding zeros as needed. The radius of Earth is about 6,380,000 m.

**Example 6:** The mass of a dust particle can be expressed as $7.53 \times 10^{-10}$ kilogram. Write the mass in standard notation.

Since the power of 10 is –10, move the decimal point ten places to the left, adding zeros as needed. The mass of a dust particle is 0.000 000 000 753 kg.

## COMPARING AND ORDERING

To put these diameters of Jupiter's moons in order from least to greatest:

| | |
|---|---|
| Io | 3630 km |
| Europa | 3138 km |
| Ganymede | 5252 km |
| Callisto | 4800 km |

proceed as follows:

| 1. Line the diameters up vertically by the ones place. | 2. Compare numbers, starting at the left column. | | 3. Rearrange by finding the first column where the numbers are different. |
|---|---|---|---|
| 3630 | 3630 | Compare **columns** | 3138 |
| 3138 | 3138 | from left to right. 5 is | 3630 |
| 5252 | **5**252 | largest, so 5252 is the | 4800 |
| 4800 | 4800 | greatest number. | 5252 |

Try ordering the decimal numbers 0.0018, 1.8, 0.018, and 0.81.

| 1. Line the numbers up vertically by the decimal point. | 2. Compare numbers, starting at the left column. | | 3. Rearrange by finding the first column where the numbers are different. |
|---|---|---|---|
| 0.0018 | 0.0018 | Compare **columns** | 0.0018 |
| 1.8 | **1**.8 | from left to right. 1 is | 0.018 |
| 0.018 | 0.018 | largest, so 1.8 is the | 0.18 |
| 0.81 | 0.81 | greatest. | 1.8 |

## DIVISIBILITY RULES

A time-saving method for many things, including putting fractions into simplest form, is the use of divisibility rules. These rules allow you to know when a number can be divided by another number.

| *You can divide by this number* | *If* |
|---|---|
| 2 | the given number ends in 0, 2, 4, 6, or 8 (it's even) |
| 3 | the sum of the given number's digits can be divided by 3 |
| 4 | the last two digits in the given number can be divided by 4 |
| 5 | the given number ends in 5 or 0 |
| 6 | the given number is divisible by both 2 and 3 |
| 9 | the sum of the given number's digits can be divided by 9 |
| 10 | the given number ends in 0 |

**Example:** Answer the following without using a calculator:

1. Is 125 divisible by 5? Yes, because 125 ends in 5.
2. Is 135 divisible by 9? Yes, because 1 + 3 + 5 = 9 and 9 can be divided by 9.
3. Is 148 divisible by 3? No, because 1 + 4 + 8 = 13 and 13 cannot be divided by 3.
4. Is 4028 divisible by 4? Yes, because the last two digits, 28, can be divided by 4.

5. Is 90,324 divisible by 6? Yes, because 90,324 can be divided by both 2 and 3.
6. Is 1098 divisible by 2? Yes, because 1098 ends in an even number.

## FACTORS

*Factors* are numbers that can be multiplied together to get a product.

**Example:** The factors of 24 are:
$1 \times 24 = 24$
$2 \times 12 = 24$
$3 \times 8 = 24$
$4 \times 6 = 24$

Therefore, 1, 2, 3, 4, 6, 8, 12, and 24 are all factors of 24.

Another name for 1, 2, 3, 4, 6, 8, 12, and 24 is *divisors* of 24, because these numbers all divide evenly into 24.

## PRIMES, COMPOSITES, AND FACTOR TREES

*Prime numbers* can be divided by only themselves and 1. The number 1 is **not** prime.

**Example 1:** Is 13 a prime number?

Yes. The only factors of 13 are 13 and 1. These two numbers are the only two that divide evenly into 13.

**Example 2:** Is 35 a prime number?

No. The factors of 35 are 1, 5, 7, and 35. Therefore, 35 is not a prime number. It is a *composite number.*

*Composite numbers* have more than two factors. They can be divided by numbers other than themselves and 1.

**Example 1:** Is 23 a composite number?

No. The only factors of 23 are 23 and 1. 23 is a prime number.

**Example 2:** Is 99 a composite number?

Yes. 99 can be divided by 3, 9, 11, 33, and 99.

Composite numbers can be broken down into prime factors. This is easy to do using a *factor tree.* Break each composite number down into its prime factors until there are no more composite numbers.

$$\begin{array}{c} 24 \\ 2 \times 12 \\ 2 \times 2 \times 6 \\ 2 \times 2 \times 2 \times 3 \end{array}$$

The prime factors of 24 are in the last row.

**Example:** Use a factor tree to express 120 as a product of prime factors.

$$\begin{array}{c} 120 \\ 12 \times 10 \\ 2 \times 6 \times 2 \times 5 \\ 2 \times 2 \times 3 \times 2 \times 5 \end{array}$$

Because product means multiply, the prime factors of 120 are $2 \times 2 \times 3 \times 2 \times 5$. These factors can also be written, using powers, as $2^3 \times 3 \times 5$.

## ABSOLUTE VALUE

The *absolute value* of a number represents its distance from 0 on the number line. Absolute value is represented by placing bars around the number. For example, the absolute value of negative 6 is shown as $|-6|$. Since negative 6 is exactly six spaces from 0 on the number line, we say $|-6| = 6$. In fact, the absolute value of a number is positive whether the number is positive or negative.

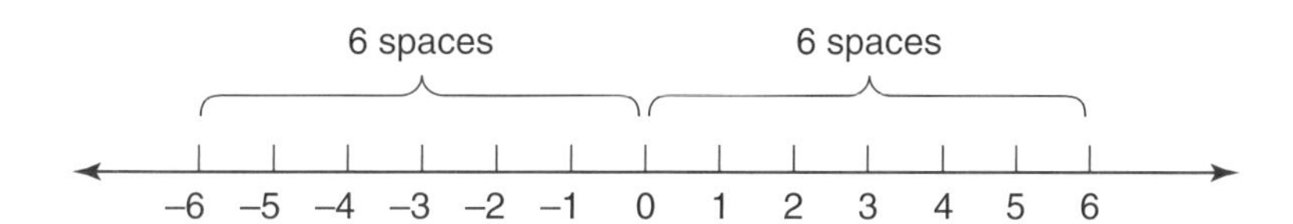


**Examples:**

1. $|1.4| = 1.4$
2. $|-3| = 3$
3. $|1 + 2| = |3| = 3$
4. $|1 - 2| = |-1| = 1$
5. $|-3 \cdot 5| = |-15| = 15$
6. $4|7| = 4 \cdot 7 = 28$
7. $-|10| = -10$ (This is read as "the opposite of the absolute value of 10.")
8. $-|-5| = -5$ (the opposite of the absolute value of negative 5)
9. $|20| - |15| = 5$
10. $4|-2| = 4 \cdot 2 = 8$

# SAMPLE QUESTIONS

## Place Value and Rounding

1. What is the value of the 3 in 2341?

   A. 3000  
   B. 300  
   C. 30  
   D. 3

2. Write four thousand three hundred as a number in standard form.

   F. 40,030  
   G. 4300  
   H. 4030  
   I. 400,300

3. Write five hundred fifty-four thousand two hundred eleven as a number in standard form.

   A. 554,200,011  
   B. 554,211  
   C. 54,211  
   D. 554.211

4. Write 1401 in words.

   F. one thousand four hundred and one  
   G. one thousand four hundred one  
   H. one thousand forty-one  
   I. fourteen thousand one

5. Write 123.002 in words.

   A. one hundred twenty-three thousand and two  
   B. one hundred twenty-three thousand two  
   C. one hundred twenty-three and two thousandths  
   D. one hundred twenty-three and two hundredths

6. Round 10,256 to the hundreds place.

7. What do all the numbers in the circle have in common?

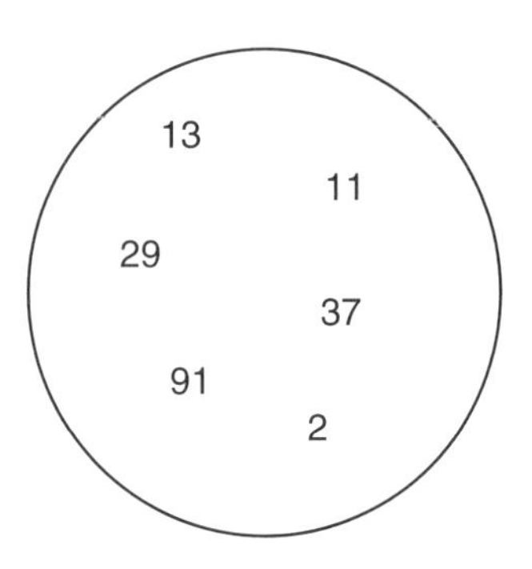


   F. All the numbers are greater than 2.  
   G. All the numbers are odd.  
   H. All the numbers are prime.  
   I. All the numbers are composite.

8. Round 0.5249 to the thousandths place.

9. Select the best name for –1.

   A. natural number  
   B. whole number  
   C. integer  
   D. odd number

10. What is the value of the 5 in 6125?

11. Write 24,310 in expanded notation.

   F. $(2 \times 10{,}000) + (4 \times 1000) + (3 \times 100) + (1 \times 10)$  
   G. $(2 \times 10{,}000) + (4 \times 100) + (3 \times 10) + (1 \times 1)$  
   H. $(24 \times 1000) + (310 \times 100)$  
   I. $(2 \times 10) + (4 \times 100) + (3 \times 1000) + (1 \times 10{,}000)$

12. Which of the following represents 0.000152 in scientific notation?

A. $152 \times 10^{-6}$
B. $152 \times 10^{4}$
C. $1.52 \times 10^{-4}$
D. $15.2 \times 10^{5}$

13. The diameter of Jupiter can be expressed in scientific notation as approximately $1.43 \times 10^{5}$ kilometers (km). Which of the following represents $1.43 \times 10^{5}$ in standard notation?

F. 143,000,000 km
G. 14,300,000 km
H. 1,430,000 km
I. 143,000 km

14. The number 12,345 is divisible by which of the following numbers?

A. 2 and 3
B. 3, 5, and 6
C. 3 and 5
D. 2, 3, 5

15. Express 75 as a product of prime factors

F. $5 \times 5 \times 5$
G. $25 \times 3$
H. $2 \times 5 \times 5$
I. $3 \times 5 \times 5$

16. A space shuttle travels in orbit at a speed of about $4.7 \times 10^{5}$ meters per hour. At this speed, how far, in meters, will the shuttle travel in 1 day?

A. $1.128 \times 10^{7}$
B. $1.128 \times 10^{8}$
C. $4.71 \times 10^{24}$
D. $4.71 \times 10^{9}$

17. What is the value of the expression $8^4 - 5^3$?

18. Which point on the number line below could represent the result of multiplying 0.1 by a number between 1 and 2?

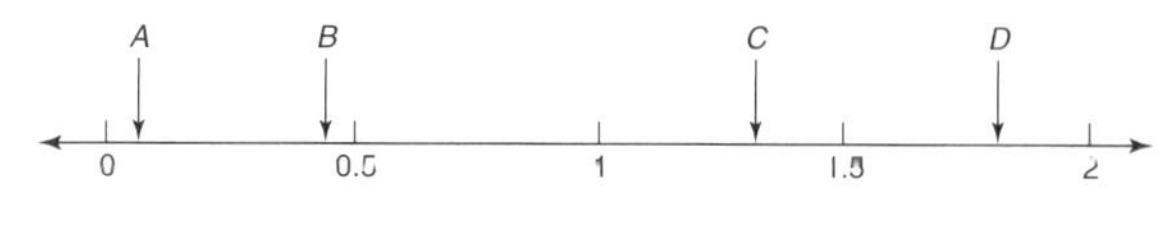


F. *A*
G. *B*
H. *C*
I. *D*

19. What is the value of the 7 in 401.4597 in the spreadsheet?

| | A | B |
|---|---|---|
| 1 | 401.4597 | |
| 2 | | |
| 3 | | |
| 4 | | |

A. 7 ten thousandths
B. 7 thousandths
C. 7 hundredths
D. 7 tenths

20. Jenna recorded the temperature at the same time every day for 1 week with the following results:

| | |
|---|---|
| Monday | 12°F |
| Tuesday | –11°F |
| Wednesday | –2°F |
| Thursday | –8°F |
| Friday | –5°F |
| Saturday | –6°F |
| Sunday | –10°F |

On which day was the temperature 6° cooler than the value of Friday's temperature?

F. Sunday
G. Monday
H. Tuesday
I. Wednesday

## ANSWERS TO SAMPLE QUESTIONS

1. **B.** Because the 3 is in the hundreds place, it has a value of 300.

2. **G.** $4 \times 1000 + 3 \times 100 = 4300$

3. **B.** 554,211

4. **G.** One thousand four hundred one. Use *and* only when there is a decimal point in the number.

5. **C.** One hundred twenty-three and two thousandths.

6. **10,300.** The 2 is in the hundreds place. The number following the 2 is 5 or greater, so round 2 to 3 and change everything after to 0.

7. **H.** All the numbers shown are prime numbers.

8. **0.525.** The 4 is in the thousandths place. The number following the 4 is more than 5, so round the 4 to 5 and drop all numbers to the right.

9. **C.** The natural numbers are counting numbers: 1, 2, 3, 4, . . . ; whole numbers: 0, 1, 2, 3, . . . ; and integers: . . . –3, –2, –1, 0, 1, 2, . . . .

10. **5.** The 5 is in the ones position and has a value of $5 \times 1$ or 5.

11. **F.** $(2 \times 10{,}000) + (4 \times 1000) + (3 \times 100) + (1 \times 10)$

12. **C.** $1.52 \times 10^{-4}$. The decimal point moves between 1 and 5. Because the decimal moved four places, you should multiply by 10 to the negative fourth power.

13. **I.** 143,000 km. The number $1.43 \times 10^5$ tells you to use 1.43 and move the decimal point five places to the right (positive power of 10), adding zeros as needed.

14. **C.** 3 and 5. The sum of the digits is 15, which is divisible by 3. Since 12,345 ends in 5, it is also divisible by 5.

15. **I.** $3 \times 5 \times 5 = 75$

16. **A.** $4.7 \times 10^5 \times 24 = 11{,}280{,}000 = 1.128 \times 10^7$

17. **3971.** $8 \times 8 \times 8 \times 8 = 4096$, and $5 \times 5 \times 5 = 125$. $4096 - 125 = 3971$.

18. **F.** Multiplying by 0.1 results in a number that is 10 times smaller than the original. For example, $1.9 \times 0.1 = 0.19$, which is much closer to 0 than points *B*, *C*, or *D*.

19. **A.** 7 ten thousandths

20. **H.** Tuesday was 6° cooler. Wednesday was 3° warmer.

# Lesson 2

# Operations on Numbers

## BASIC OPERATIONS

During Hurricane Charley, 14 trees were blown over on Orlando Street, 6 on Adorn Street, 10 on Easy Street, and 4 on Conroe Street. What was the total number of trees that were blown over?

The key word is *total*, which means add: $14 + 6 + 10 + 4 = 34$.

There are four *basic operations* on numbers: addition, subtraction, multiplication, and division.

### Addition of Integers

When you *add* integers, the result is called a *sum*. The rules for adding integers are as follows:

**1.** When adding two numbers with the same sign, simply find the sum. The sum will have the same sign.

**Examples:**

1. $5 + 2 = 7$
2. $-5 + (-3) = -8$

Key words for addition: sum, total, increased by, more than, plus

**2.** When adding two numbers with different signs, subtract the absolute value of one number from that of the other. The sum will have the sign of the number with the greater absolute value.

**Examples:**

1. $-7 + 2 = -5$
2. $8 + (-1) = 7$

## Subtraction of Integers

When you *subtract* integers, the result is called a *difference*. If you are asked to find the difference of two numbers, you need to subtract.

To subtract integers, change the subtraction sign to addition, and then change the sign of the number being subtracted (the second number). After that, follow the same rules as for addition.

### Examples:

1. 10 – 7 is the same as 10 + (–7). The subtraction sign changed to addition, and 7 changed to –7. The answer is 3.
2. 7 – 10 is the same as 7 + (–10). The subtraction sign changed to addition, and 10 changed to –10. The answer is –3.
3. –4 – 6 is the same as –4 + (–6). The subtraction sign changed to addition, and 6 changed to –6. The answer is –10.
4. –4 – (–6) is the same as –4 + 6. The subtraction sign changed to addition, and –6 changed to 6. The answer is 2.

| Key words for subtraction: | difference | minus |
|---|---|---|
| | less | decreased by |
| | less than | |

**Example:** Andrea earned \$120 less than Earl last week. If Earl earned \$350, how much did Andrea earn?

The key words are *less than*. Subtract \$120 from \$350. \$350 – \$120 = \$230.

## Multiplication of Integers

When you *multiply* integers, the result is called a *product*. The product of 4 and 3 is 12. On the FCAT you may see different ways of representing multiplication:

| | |
|---|---|
| Multiplication (times) sign: | $4 \times 3$ |
| A dot: | $4 \cdot 3$ |
| Parentheses or brackets: | 4(3) or 4[3] or (4)(3) |
| Absolute-value bars: | 4\|3\| |
| No symbol: | 3*a* (means 3 times *a*)<br>*ab* (means *a* times *b*) |

The rules for multiplying integers are as follows:

| Rule | Example |
|---|---|
| positive · positive = positive | $10 \cdot 2 = 20$ |
| positive · negative = negative | $10 \cdot -2 = -20$ |
| negative · negative = positive | $-10 \cdot -2 = 20$ |
| negative · positive = negative | $-10 \cdot 2 = -20$ |

Key words for multiplication: product, times, multiplied by, of

**Example 1:** Fred ate $\frac{1}{3}$ of a pizza. If there were 12 slices, how many did he eat?

The key word is *of*. Multiply 12 by $\frac{1}{3}$. $\frac{12}{1} \cdot \frac{1}{3} = \frac{12}{3} = 4$. Fred ate 4 slices.

**Example 2:** Angie drove 125 miles each week for 5 weeks. What is the total number of miles she drove?

The key words are *each* and *total*. Add 125 five times, or muliply 125 by 5: $125 \cdot 5 = 625$.

## Division of Integers

When you *divide* two integers, the result is called a *quotient*. There are three common ways of representing division:

Division symbol: $9 \div 3$

Long division symbol: $3\overline{)9}$

Fraction bar: $\frac{9}{3}$

The rules for dividing integers are the same as those for multiplying except for the change in operation.

| Rule | Example |
|---|---|
| positive ÷ positive = positive | $10 \div 2 = 5$ |
| positive ÷ negative = negative | $10 \div -2 = -5$ |
| negative ÷ negative = positive | $-10 \div -2 = 5$ |
| negative ÷ positive = negative | $-10 \div 2 = -5$ |

**Important! Numbers cannot be divided by zero. We say that division by zero is *undefined*. For example, $\frac{5}{0}$ is undefined, but $\frac{0}{5} = 0$.**

Key words for division: quotient
divided by
words that indicate a total amount is to be split evenly
per (as in "miles per gallon")

**Example:** Five friends went out to dinner at a fancy restaurant. The total bill was \$632.55. If the bill is to be split evenly, how much should each person pay?

The key words here indicate that \$632.55 is to be *split evenly* between the friends: $\$632.55 \div 5 = \$126.51$.

## ORDER OF OPERATIONS

You may be asked to perform more than one operation at a time. It is very important that you know in what order to do operations as the calculator you will be using on the FCAT is not programmed to do operations in the correct order.

DO FIRST: All operations requiring grouping symbols. Grouping symbols are parentheses, brackets, and fraction bars. For fraction bars, do all operations on the top and bottom first; then, if possible, divide.

DO NEXT: All operations with exponents.

DO NEXT: All operations with either division or multiplication. Be sure to work from left to right.

DO LAST: All operations with either addition or subtraction. Be sure to work from left to right.

**Example 1:** Find the value of $4 - 2(6 + 3)$.

| | | |
|---|---|---|
| DO FIRST: | $4 - 2(9)$ | parentheses |
| DO NEXT: | $4 - 18$ | multiply |
| DO LAST: | $-14$ | subtract |

**Example 2:** Find the value of $\frac{5+3^2}{7}$.

| | | |
|---|---|---|
| DO FIRST: | $\frac{5+9}{7}$ | exponent |
| DO NEXT: | $\frac{14}{7}$ | add |
| DO LAST: | 2 | fraction bar (divide) |

**Example 3:** Which operation should be performed first to find the value of $12 + 6 \div 2 \times 3$?

Do the division first: $6 \div 2 = 3$. Then $12 + 3 \times 3 = 12 + 9 = 21$. Remember: when doing division or multiplication, work from left to right.

**Example 4:** Simplify $5(1 + 2)^2 - 20 \div 5$.

| | | |
|---|---|---|
| DO FIRST: | $5(3)^2 - 20 \div 5$ | parentheses |
| DO NEXT: | $5(9) - 20 \div 5$ | exponent |
| DO NEXT: | $45 - 20 \div 5$ | multiplication (left to right) |
| DO NEXT: | $45 - 4$ | division |
| DO LAST: | 41 | subtraction |

# SAMPLE QUESTIONS

## Operations on Numbers

In 1–13, simplify each expression:

1. $-9 + 6$

2. $14 - 22$

3. $-3(5)$

4. $\frac{2-17}{5}$

5. $4(6 - 8)$

6. $4 + 9 \div 3 + 1$

7. $5 \cdot 4 + 3 \cdot 5$

8. $2 + 4^2 - 10$

9. $3(6 - 8)^3 \div 6 + 1$

10. $\frac{2(3 \cdot 2^3)}{12}$

11. $\frac{4(5-2)}{6} + 4$

12. $\frac{1}{2}(4 + 6) - 5$

13. $-1 \cdot 5 + 3 \cdot 0$

14. If 8 is divided by $\frac{1}{2}$ and 2 is added to the quotient, the result is:

A. 6
B. 12
C. 18
D. 20

15. During the winter, the temperature in Fairbanks, Alaska, fell to 55 degrees below zero.

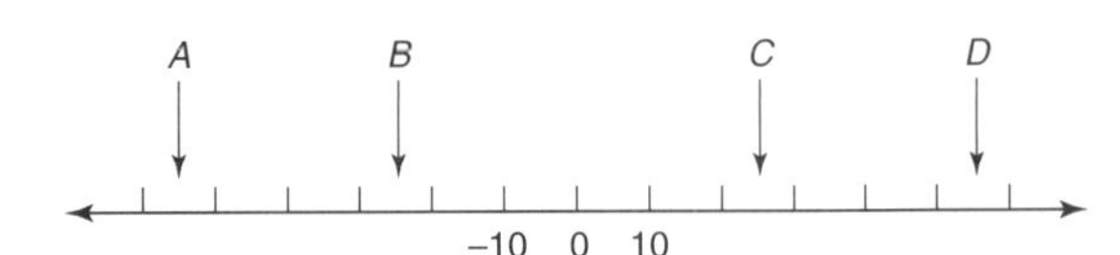


Which letter on the number line represents that temperature?

F. *A*
G. *C*
H. *B*
I. *D*

16. Gary worked 50 hours last week. He gets paid overtime for any hours over 40 that he works. Overtime pay is 1.5 times regular pay. If Gary earns \$9 an hour, how much was his total pay for last week?

17. Mr. Schwartz bought 440 gallons of gasoline last year. If the average price of gasoline was \$1.82 per gallon, how much did he spend on gas?

18. If Mr. Schwartz drove 10,340 miles last year, how many miles per gallon did he average if he bought 440 gallons of gas?

19. Jay is shopping for a new garage door. One estimate for the door is $1250 if cash is paid. A second estimate offers the same door for $450 down and $75 per month for 12 months. How much will Jay save by paying cash?

20. Andrea has to read a 567-page book during the first 9 weeks of school. If she read 207 pages during the first week, how many pages per week does she need to read in order to finish by the end of 9 weeks?

## ANSWERS TO SAMPLE QUESTIONS

1. **–3.**

2. **–8.** $14 - 22 = 14 + (-22) = -8$

3. **–15.**

4. **–3.** $\frac{2-17}{5} = \frac{2+(-17)}{5} = \frac{-15}{5} = -3$

5. **–8.** $4(6 - 8) = 4(6 + (-8)) = 4(-2) = -8$

6. **8.** $4 + 9 \div 3 + 1 = 4 + 3 + 1 = 8$

7. **35.** $5 \cdot 4 + 3 \cdot 5 = 20 + 15 = 35$

8. **8.** $2 + 4^2 - 10 = 2 + 16 - 10 = 18 - 10 = 8$

9. **–3.** $3(6 - 8)^3 \div 6 + 1 = 3(-2)^3 \div 6 + 1 = 3(-8) \div 6 + 1 = -24 \div 6 + 1 = -4 + 1 = 3$

10. **4.** $\frac{2(3 \cdot 2^3)}{12} = \frac{2(3 \cdot 8)}{12} = \frac{2(24)}{12} = \frac{48}{12} = 4$

11. **6** $\frac{4(5-2)}{6} + 4 = \frac{4(3)}{6} + 4 = \frac{12}{6} + 4 = 2 + 4 = 6$

12. **0.** $\frac{1}{2}(4 + 6) - 5 = \frac{1}{2}(10) - 5 = \frac{1}{2} \cdot \frac{10}{1} - 5 = 5 - 5 = 0$

13. **–5.** $-1 \cdot 5 + 3 \cdot 0 = -5 + 0 = -5$

14. **C.** $8 \div \frac{1}{2} + 2 = \frac{8}{1} \div \frac{1}{2} + 2 = \frac{8}{1} \cdot \frac{2}{1} + 2 = 16 + 2 = 18$

15. **F.**

16. **$495.** This is a multistep problem.
Step 1: Subtract 40 from 50 to find out how many overtime hours Gary worked: 50 – 40 = 10.
Step 2: Multiply 10 × 1.5 × 9 to find out how much pay he received for the overtime: $135.
Step 3: Multiply 40 × $9 to find the regular pay: $360.
Step 4: Add the overtime pay to the regular pay ($360 + $135) to get the total pay: $495.

17. **$800.80.** Multiply 440 by $1.82.

18. **23.5.** Divide 10,340 miles by 440 gallons to get miles per gallon.

19. **$100.** This is a multistep problem.
Step 1: Multiply $75 by 12 to find the annual cost if cash is not paid: $900.
Step 2: Add the down payment of $450 to $900: $1350.
Step 3: Subtract $1250 from $1350 to find the difference: $100.

20. **45.** This is a multistep problem.
Step 1: Subtract 207 from 567 to find how many pages Andrea still needs to read: 360.
Step 2: Divide 360 by 8 (she has 8 weeks left to read these pages): 45.

# Lesson 3

## Fractions

If you slice a pizza into 8 equal parts and you eat 5 of them, what fractional part of the pizza have you eaten?

If you ate 5 parts out of a total of 8, the part that you ate can be represented by the fraction $\frac{5}{8}$.

Fractions are used to represent parts of a whole. The bottom number of the fraction is called the *denominator* and represents the whole. The top number of the fraction is called the *numerator* and represents the part.

### PROPER AND IMPROPER FRACTIONS

When the numerator is smaller than the denominator, the fraction is said to be a *proper fraction*. When the numerator is larger than the denominator, the fraction is called an *improper fraction*. Improper fractions represent amounts that are larger than 1.

**Examples of proper fractions:** $\frac{1}{2}, \frac{3}{4}, \frac{4}{5}, \frac{2}{3}$

**Examples of improper fractions:** $\frac{5}{2}, \frac{7}{4}, \frac{12}{5}, \frac{4}{3}$

### MIXED NUMBERS

A *mixed number* contains both a whole number and a fraction.

**Examples of mixed numbers:** $1\frac{2}{5}, 3\frac{5}{9}, 5\frac{1}{2}$

### CHANGING MIXED NUMBERS TO FRACTIONS

To change a mixed number to a fraction, multiply the whole number by the denominator of the fraction and then add the numerator. The result is the numerator of the new fraction. Keep the old denominator.

This is an important skill for the FCAT because you cannot grid a mixed number. Mixed numbers must be changed to improper fractions or decimal forms.

**Example:** Change $1\frac{2}{5}$ to an improper fraction.

Multiply the whole number by the denominator: $1 \times 5 = 5$
Add the numerator: $5 + 2 = 7$

The new numerator is 7.
The old denominator was 5.

The mixed number $1\frac{2}{5}$ equals $\frac{7}{5}$.
This should be gridded as 7/5 and justified all the way to the right or all the way to the left. On the FCAT a machine scores your response and may score it incorrectly if it runs into blank spaces in the first and last column.

## CHANGING IMPROPER FRACTIONS TO MIXED NUMBERS

When changing an improper fraction to a mixed number, divide the numerator by the denominator. The quotient is the whole number, and the remainder becomes the new numerator. As before, keep the old denominator.

**Example:** Change $\frac{9}{5}$ to a mixed number.

$$\begin{array}{r} 1 \\ 5\overline{)9} \\ \underline{-5} \\ 4 \end{array}$$ (remainder)

Mixed number = $1\frac{4}{5}$

## EQUIVALENT FRACTIONS

Fractions that represent the same amount are called *equivalent fractions.*

**Examples:** $\frac{1}{2}, \frac{2}{4}, \frac{3}{6}, \frac{4}{8}$ are all equivalent fractions. One way to tell whether they are equivalent is to divide the numerator of each fraction by the denominator. For this group of fractions, the result is always 0.5.

Another way to tell whether two fractions are equivalent is to cross-multiply them.

**Example 1:** Is $\frac{4}{5}$ equivalent to $\frac{8}{10}$?

$40 = 40$

$\frac{4}{5} \times \frac{8}{10}$

The cross products are the same (40), so the fractions are equivalent.

**Example 2:** Is $\frac{3}{5}$ equivalent to $\frac{6}{12}$?

$36 \neq 30$

$\frac{3}{5} \times \frac{6}{12}$

The cross products are not the same; therefore, $\frac{3}{5}$ is not equivalent to $\frac{6}{12}$.

In fact, because 36 is more than 30, $\frac{3}{5}$ is greater than $\frac{6}{12}$.

## PUTTING FRACTIONS INTO LOWEST TERMS

A fraction is not in lowest terms until the numerator and the denominator share no common factors.

**Example 1:** Is $\frac{5}{10}$ in lowest terms?

The fraction $\frac{5}{10}$ is not in lowest terms because 5 and 10 share a common factor of 5. To put $\frac{5}{10}$ into lowest terms, simply divide both the numerator and the denominator by 5. The resulting fraction, $\frac{1}{2}$, is in lowest terms.

**Example 2:** Mr. Ott has harvested 36 acres of his 48-acre orange grove. Express the harvested area as a fraction in lowest terms.

Mr. Ott has harvested $\frac{36}{48}$ of his orange grove. The numerator (36) and the denominator (48) share a common factor of 12. Divide by 12. The resulting fraction, $\frac{3}{4}$, is equivalent to $\frac{36}{48}$ and is in lowest terms.

# ADDING AND SUBTRACTING FRACTIONS AND MIXED NUMBERS

## Fractions

To add or subtract fractions with the same denominator, simply add (or subtract) the numerators. Keep the denominator the same.

**Example 1:** Add $\frac{2}{5} + \frac{1}{5}$.

$$\frac{2}{5} + \frac{1}{5} = \frac{3}{5}$$

**Example 2:** What is the value of $\frac{7}{8} - \frac{4}{8}$?

$$\frac{7}{8} - \frac{4}{8} = \frac{3}{8}$$

There are many methods for adding and subtracting fractions with unlike denominators. If you are accomplished at putting fractions into lowest terms, the following method may be ideal for you.

**Example 1:** Add $\frac{5}{8} + \frac{1}{4}$.

Multiply the denominators to get a new common denominator: 32.
Cross-multiply the fractions, and **add** the results to get the new numerator:

20 + 8

$$\frac{5}{8} \times \frac{1}{4} = \frac{28}{32}$$

Put $\frac{28}{32}$ into lowest terms (divide by the common factor, 4): $\frac{28}{32} = \frac{7}{8}$.

**Example 2:** Subtract $\frac{11}{12} - \frac{2}{5}$.

Multiply the denominators to get a common denominator: 60.
Cross-multiply the fractions, and **subtract** the results to get the new numerator:

55 – 24

$$\frac{11}{12} \times \frac{2}{5} = \frac{31}{60}$$

Since $\frac{31}{60}$ is already in lowest terms, this is your answer.

## Mixed Numbers

You can use a similar technique to add and subtract mixed numbers. First, change the mixed numbers to improper fractions; then use the methods for adding and subtracting shown above.

**Example 1:** Add $3\frac{3}{5} + 2\frac{3}{8}$.

Change the mixed numbers to improper fractions: $\frac{18}{5} + \frac{19}{8}$.

Multiply the denominators to get a common denominator: 40. Cross-multiply the fractions, and **add** the results to get the new numerator.

144 + 45

$$\frac{18}{5} \times \frac{9}{8} = \frac{189}{40}$$

Change back to a mixed number by dividing the numerator by the denominator and using the remainder as your new numerator:

$\frac{189}{40} = 4\frac{29}{40}$.

**Example 2:** Subtract $2\frac{7}{8} - 1\frac{4}{5}$.

Change the mixed numbers to improper fractions: $\frac{23}{8} - \frac{9}{5}$.

Multiply the denominators to get a common denominator: 40. Cross-multiply the fractions, and **subtract** the results to get the new numerator.

115 – 72

$$\frac{23}{8} \times \frac{9}{5} = \frac{43}{40}$$

Change back to a mixed number by dividing the numerator by the denominator and using the remainder as your new numerator:

$\frac{43}{40} = 1\frac{3}{40}$.

## MULTIPLYING AND DIVIDING FRACTIONS AND MIXED NUMBERS

### Fractions

Multiplying fractions is particularly easy. Simply multiply the numerators and then the denominators. Put into lowest terms if needed.

**Example:** Multiply $\frac{3}{4} \times \frac{4}{5}$.

Multiply the numerators and denominators: $\frac{12}{20}$.

Express in lowest terms: $\frac{12}{20} = \frac{3}{5}$.

To divide fractions, invert the **second** fraction and then multiply.

**Example:** Divide $\frac{2}{5} \div \frac{1}{3}$.

$\frac{2}{5} \times \frac{3}{1}$ Invert the **second** fraction.

$\frac{6}{5}$ Multiply the numerators and denominators.

### Mixed Numbers

To multiply or divide mixed numbers, it is necessary to change each mixed number to an improper fraction before multiplying or dividing.

**Example 1:** Multiply $1\frac{2}{3} \times 2\frac{3}{5}$.

$\frac{5}{3} \times \frac{13}{5}$ Change to improper fractions.

$\frac{65}{15}$ Multiply the numerators and denominators:

$\frac{13}{3}$ Put into lowest terms.

**Example 2:** Divide $2\frac{4}{5} \div \frac{1}{2}$.

$\frac{14}{5} \div \frac{1}{2}$ Change the mixed number to an improper fraction.

$\frac{14}{5} \times \frac{2}{1}$ Invert the second fraction.

$\frac{28}{5}$ Multiply the numerators and denominators.

**Example 3:** Divide $5\frac{1}{4}$ by 10.

$\frac{21}{4} \div \frac{10}{1}$ Change the mixed number and whole number to improper fractions.

$\frac{21}{4} \times \frac{1}{10}$ Invert the second fraction.

$\frac{21}{40}$ Multiply the numerators and denominators.

## CHANGING DECIMAL NUMBERS TO FRACTIONS AND FRACTIONS TO DECIMAL NUMBERS

Sometimes it is necessary to compare numbers that are in different forms. When this happens, it is helpful to put the numbers into the same form.

**Example:** Which is larger, 0.15 or $\frac{1}{5}$?

To compare the decimal to the fraction, first change the decimal into fraction form. Note that 0.15 is 15 hundredths, which, in fraction form, is represented as $\frac{15}{100}$, or $\frac{3}{20}$ in lowest terms. Now use cross-multiplication to compare the fractions:

15 20

$\frac{3}{20} \times \frac{1}{5}$

Since 20 times 1 is larger than 5 times 3, $\frac{1}{5}$ is the larger number.

A second way to compare the two numbers is to leave 0.15 as a decimal and change $\frac{1}{5}$ into decimal form. Simply divide the numerator by the denominator; thus $1 \div 5 = 0.2$. Since 0.2 is larger than 0.15 (comparing 0.20 to 0.15), the fraction $\frac{1}{5}$ is larger than .15.

## COMPARING GROUPS OF FRACTIONS

When comparing or ordering a group of fractions with unlike denominators, you may find it helpful to change each fraction to a decimal first.

**Example:** Sharda kept this record of her math quizzes for the first 4 weeks of school:

| Quiz | Fraction Answered Correctly |
|---|---|
| 1 | $\frac{5}{7}$ |
| 2 | $\frac{5}{6}$ |
| 3 | $\frac{7}{8}$ |
| 4 | $\frac{6}{7}$ |

In which quiz did she answer the largest fraction of questions correctly?

Change each fraction to a decimal as shown in the table below. Then compare the decimals to see that 0.88 is the largest. Thus, the fraction $\frac{7}{8}$ represents the largest fraction of correct answers.

| Quiz | Fraction Answered Correctly |
|---|---|
| 1 | $\frac{5}{7} = 0.71$ |
| 2 | $\frac{5}{6} = 0.83$ |
| 3 | $\frac{7}{8} = 0.88$ |
| 4 | $\frac{6}{7} = 0.86$ |

# SAMPLE QUESTIONS

## Fractions

1. When Frank's party was over, he had $1\frac{3}{8}$ pizzas left. Express this number as an improper fraction.

2. Which of the following is NOT equivalent to $\frac{4}{5}$?

   A. $\frac{8}{10}$
   B. $\frac{12}{15}$
   C. $\frac{8}{12}$
   D. $\frac{16}{20}$

3. Find the sum of $\frac{5}{6}$ and $\frac{2}{3}$. Express your answer as a fraction in lowest terms.

4. What is the value of $2\frac{1}{2} - \frac{4}{5}$?

5. If seven candy bars are divided among 21 people, how much candy will each person get?

6. Chandra needs to mix $1\frac{1}{3}$ cups of flour with $\frac{3}{4}$ cups of sugar. Her mixing bowl holds exactly 2 cups. Is her bowl large enough to hold both the flour and the sugar? If not, how much larger does the bowl need to be?

7. The following are some common wrench measurements. Which set is arranged in ascending (smallest to largest) order?

   F. $\frac{3}{8}, \frac{5}{8}, \frac{3}{16}, \frac{1}{4}$
   G. $\frac{3}{16}, \frac{5}{8}, \frac{1}{4}, \frac{3}{8}$
   H. $\frac{3}{16}, \frac{5}{8}, \frac{3}{8}, \frac{1}{4}$
   I. $\frac{3}{16}, \frac{1}{4}, \frac{3}{8}, \frac{5}{8}$

8. When a fraction less than 1 is multiplied by a whole number greater than 1, which of the following could be the result?

   A. a number larger than the whole number
   B. a number less than the whole number
   C. a number less than the fraction
   D. zero

9. Which number is closest to $\frac{5}{6}$?

   F. 0.8
   G. 0.83
   H. 0.5
   I. 0.16

**10.** The diagonals in the cube represent equivalent fractions.

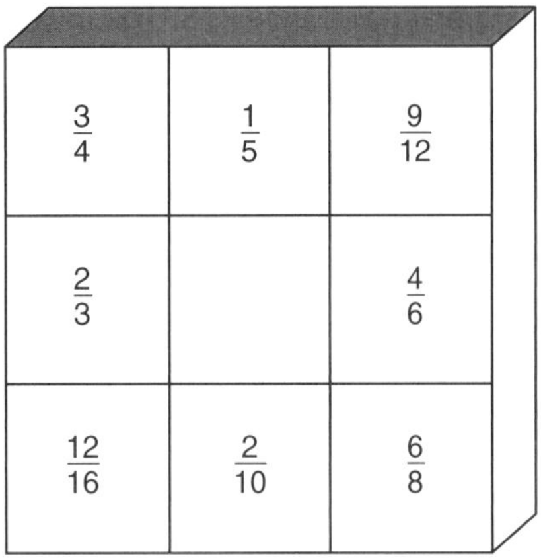

| | | |
|---|---|---|
| $\frac{3}{4}$ | $\frac{1}{5}$ | $\frac{9}{12}$ |
| $\frac{2}{3}$ | | $\frac{4}{6}$ |
| $\frac{12}{16}$ | $\frac{2}{10}$ | $\frac{6}{8}$ |

Which fraction could be the fraction missing from the center?

A. $\frac{6}{12}$
C. $\frac{15}{20}$
B. $\frac{9}{15}$
D. $\frac{12}{20}$

**11.** Jay's watering can holds 2 gallons of water. He needs to water 24 plants, and each will require $\frac{1}{2}$ gallon of water. How many times will Jay need to refill the watering can?

**12.** Two fractions each less than 1 are multiplied together. Their product will be

F. larger than either fraction
G. smaller than either fraction
H. larger than one fraction and smaller than the other
I. always equal to 1

**13.** Parts of two cookie recipes are shown below.

| Ingredient | Recipe 1 | Recipe 2 |
|---|---|---|
| Flour | 2 cups | 3 cups |
| Sugar | 1 cup | $1\frac{1}{2}$ cups |
| Butter | 1 cup | 1 cup |

Which recipe calls for the greater ratio of sugar to the total recipe?

A. The two recipes require equal ratios
B. Recipe 1
C. Recipe 2
D. You cannot tell from the information given.

**14.** Multiplying which of the following numbers by itself will give an answer that is larger than that number?

F. $\frac{9}{10}$
H. .875
G. 0.2
I. $\frac{3}{2}$

**15.** Each share of Angie's stock has increased in value by $\frac{3}{8}$. If she bought 100 shares at $\$23\frac{1}{4}$ per share, how much money has she made from this increase?

**16.** Between what two consecutive integers is $\frac{4}{3}$ found?

**A.** 0 and 1 **C.** 1 and 2
**B.** 2 and 3 **D.** 3 and 4

**17.** Between what two consecutive integers is $-\frac{7}{2}$ found?

**F.** –3 and –4 **H.** –2 and –3
**G.** –1 and –2 **I.** 0 and –1

18. Raven is working with her father to make a bookcase. She measures the space available and decides that each shelf should be $2\frac{3}{4}$ feet long. If the bookcase has four shelves, will a 12-foot board be enough to make the shelf? If not, how much more or less will be needed?

**19.** Hank runs 5 miles per day. He has already run $3\frac{2}{3}$ miles. How many miles does he have left to run?

**20.** If the Cookie Monster eats $\frac{1}{2}$ of a cookie on the first day, $\frac{1}{4}$ of the cookie on the second day, and $\frac{1}{8}$ of the cookie on the third day, how much of the cookie is left after the third day?

## ANSWERS TO SAMPLE QUESTIONS

1. **$\frac{11}{8}$.** Multiply: $1 \times 8 + 3 = 11$. Keep the same denominator.

2. C.

3. **$\frac{3}{2}$.** $\frac{5}{6} + \frac{2}{3} = \frac{27}{18} = \frac{3}{2}$

4. **$\frac{17}{10}$.** $2\frac{1}{2} - \frac{4}{5} = \frac{5}{2} - \frac{4}{5} = \frac{25-8}{10} = \frac{17}{10}$. Another way to do this problem is to change both fractions to decimals: $2.5 - 0.8 = 1.7$; 1.7 and $\frac{17}{10}$ are equal.

5. **$\frac{1}{3}$.** $\frac{7}{21} = \frac{1}{3}$

6. **No.** $\frac{1}{12}$. $1\frac{1}{3} + \frac{3}{4} = \frac{4}{3} + \frac{3}{4} = \frac{25}{12} = 2\frac{1}{12}$, which is $\frac{1}{12}$ more than the bowl can hold.

7. **I.** $\frac{3}{16}, \frac{1}{4}, \frac{3}{8}, \frac{5}{8}$ are arranged in order from smallest to largest. To compare the sets, you can change all the measurements to equivalent fractions (denominator of 16) or to decimals.

8. **B.** When a fraction less than 1 is multiplied by a whole number greater than 1, the result will always be smaller than the original whole number. When in doubt, use two numbers; for example, $\frac{1}{4}$ multiplied by 2 equals $\frac{1}{2}$.

9. **G.**

10. **C.** $\frac{15}{20}$ is equivalent to $\frac{3}{4}$ and to all the other fractions on the diagonals.

11. **6.** The watering can holds 2 gallons of water, which will water 4 plants; 24 divided by 4 equals 6.

12. **G.** When a fraction less than 1 is multiplied by another fraction less than 1, the result will always be smaller than either of the original fractions; for example, $\frac{1}{3}$ multiplied by $\frac{1}{2}$ equals $\frac{1}{6}$.

13. **C.** The ratio of sugar in recipe 1 is $\frac{1}{4}$ (1 cup of sugar for 4 cups of recipe). The ratio of sugar in recipe 2 is $\frac{1.5}{5.5}$ (1.5 cups of sugar for 5.5 cups of recipe). It is easy to compare the two fractions if you change them to decimal form: $\frac{1}{4} = 0.25$, and $\frac{1.5}{5.5}$ is approximately 0.27.

14. **I.** Only $\frac{3}{2}$ represents a number greater than 1. Therefore, if you multiply it by itself, you will get a larger number: $\frac{3}{2} \times \frac{3}{2} = \frac{9}{4}$.

15. **\$37.50.** Change $\frac{3}{8}$ to decimal form (0.375) and multiply by 100.

16. **C.** $\frac{4}{3} = 1\frac{1}{3}$, so it lies between 1 and 2 on the number line.

17. **F.** $-\frac{7}{2} = -3.5$, which is located between –3 and –4 on the number line.

18. **Yes. 1 ft less** will be needed. This is a multistep problem.
Step 1: Find how many feet is needed for all four shelves by multiplying:
$4 \times 2\frac{3}{4} = \frac{4}{1} \cdot \frac{11}{4} = \frac{44}{4} = 11$ ft.
Step 2: Subtract: 12 ft 11 ft = 1 ft

19. $\mathbf{\frac{4}{3}}$. Subtract: $5 - 3\frac{2}{3} = \frac{5}{1} - \frac{11}{3} = \frac{15-11}{3} = \frac{4}{3}$. Grid the answer in improper fraction form.

20. $\mathbf{\frac{1}{8}}$.
This is a multi-step problem.
Step 1: Find out how much the Cookie Monster has eaten by adding:
$\frac{1}{2} + \frac{1}{4} + \frac{1}{8} = \frac{4}{8} + \frac{2}{8} + \frac{1}{8} = \frac{7}{8}$.
Step 2: Subtract: $1 - \frac{7}{8} = \frac{1}{8}$.

# Lesson 4

# Ratios, Rates, Proportions and Scale Drawings

## RATIOS

If there are 15 boys in a classroom and 10 girls, what is the ratio, expressed in lowest terms, of girls to boys?

We say that the ratio of boys to girls is 15 to 10 or, in lowest terms, 3 to 2. Therefore, the ratio of girls to boys is 2 to 3. Ratios can also be expressed by using a colon, *2:3*, or in fraction form, $\frac{2}{3}$.

A *ratio* is a comparison of two measures or other numbers that represent essentially the same type of quantities.

**Examples:** At the local Humane Society, there are 300 cats and 90 dogs, but no other animals.

1. The ratio of cats to dogs is 300:90 or $\frac{10}{3}$.
2. The ratio of dogs to cats is 90:300 or $\frac{3}{10}$.
3. The ratio of cats to all animals is 300 to 390 or $\frac{10}{13}$.
4. The ratio of dogs to all animals is 90 to 390 or $\frac{3}{13}$.

## RATES

A *rate* compares two measures or other numbers that represent different types of quantities. One example of a rate is $14 for 3 books. This rate can be expressed as 14:3, $\frac{14}{3}$, or simply 14 to 3.

A *unit rate* is a rate in which the denominator is 1. For example, if a car burns 22 gallons of gas per mile, the rate can be written as 22:1 or $\frac{22}{1}$.

A formula on the FCAT Reference Sheet is also used for problems involving distance, rate, and time. This formula is

$$\text{Distance} = \text{Rate} \times \text{Time}$$

$$d = rt$$

If you know two parts of the formula, you can solve for the missing part.

$$\text{Time} = \frac{\text{Distance}}{\text{Rate}}$$

$$t = \frac{d}{r}$$

$$\text{Rate} = \frac{\text{Distance}}{\text{Time}}$$

$$r = \frac{d}{t}$$

**Example 1:** An airplane travels at 250 miles per hour for 5 hours. What is the distance ($d$) covered?

Use the formula: $d = rt$
Substitute for variables $r$ and $t$: $d = 250 \cdot 5 = 1250$ mi

**Example 2:** If a car covers 240 miles in 4 hours, what is its rate per hour?

Use the formula: $r = \frac{d}{t}$

Substitute for variables $d$ and $t$: $r = \frac{240}{4} = 60$ mph

**Example 3:** A train travels at 50 miles per hour. How long will the train take to cover 1000 miles?

Use the formula: $t = \frac{d}{r}$

Substitute for variables $d$ and $r$: $t = \frac{1000}{50} = 20$ hr

## PROPORTIONS

When two ratios or rates are equal, they form a *proportion*. Equivalent fractions are proportional. As you saw in Lesson 3, the cross products of equivalent fractions are equal. This is true for all proportions. If the cross products are not equal, the two ratios do not form a proportion. Cross products will also show you which is the greater rate.

**Example 1:** A person traveling 150 miles in 3 hours is traveling at the same rate as another person traveling 50 miles in 1 hour.

$\frac{150}{3} = \frac{50}{1}$ Cross multiply $(1 \times 150)$ and $(3 \times 50)$.

Notice that the two quantities are equal. Therefore these two rates form a proportion. Both people are traveling at the same rate or speed.

**Example 2:** Ann and John are painting a fence in their backyard. Ann can paint 6 fence rails in 20 minutes. John paints 8 fence rails in 30 minutes. Who is painting at the faster rate, Ann or John?

Set up the problem using a rate for each person. Compare the two rates by cross multiplying to see which is larger.

Ann John

$\frac{6 \text{ rails}}{20 \text{ min}} \stackrel{?}{=} \frac{8 \text{ rails}}{30 \text{ min}}$

Cross-multiply $(20 \times 8)$ and $(30 \times 6)$:

Ann John

180 160

$\frac{6}{20} \neq \frac{8}{30}$

The two quantities are not equal. Ann is painting fence rails at a faster rate than John.

It is possible to use proportions to solve all sorts of problems that involve rates.

**Example 1:** Jack paid \$45,000 for 18 acres of land. What was the price for 2 acres?

This multistep question asks you to find a unit rate, that is, the price for 2 acres.

Step 1: Set up the proportion so that the items match: $\frac{\$}{\text{acres}} = \frac{\$}{\text{acres}}$.

Step 2: Fill in the correct numbers for each rate: $\frac{\$45{,}000}{18 \text{ acres}} = \frac{\$d}{2 \text{ acres}}$.

Step 3: Cross-multiply: $(2 \times 45{,}000) = 18 \times d$

$90{,}000 = 18d$

Step 4: Divide both sides by 18: $5000 = d$

Jack paid \$5000 for 2 acres.

**Example 2:** Mr. Bailey drove his car 375 miles in 7.5 hours. If he continued at the same rate, how far could he drive in 10 hours?

This question asks you to find a missing amount for a rate.

Step 1: Set up a proportion so that the words match: $\frac{\text{miles}}{\text{hours}} = \frac{\text{miles}}{\text{hours}}$.

Step 2: Fill in the correct numbers for each rate: $\frac{375 \text{ mi}}{7.5 \text{ hr}} = \frac{m \text{ mi}}{10 \text{ hr}}$.

Step 3: Cross-multiply: $(10 \times 375) = 7.5 \times m$

$3750 = 7.5m$

Step 4: Divide both sides by 7.5: $500 = m$

Mr. Bailey could drive 500 mi in 10 hr.

**Example 3:** Raven drove from Hennesey to Smallville in 5 hours. If she continues to drive at the same speed, how long will she take to go from Smallville to Wildwood?

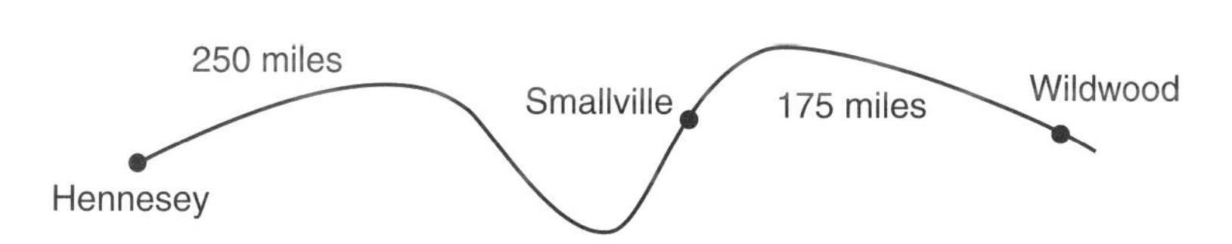

Set up a proportion: $\frac{250 \text{ mi}}{5 \text{ hr}} = \frac{175 \text{ mi}}{x \text{ hr}}$

$$175 \times 5 = 250x$$
$$875 = 250x$$
$$3.5 = x$$

Raven would take 3.5 hr to go from Smallville to Wildwood.

## MEASUREMENT CONVERSIONS

There are two methods for converting measurements from one form to another. In the first method you multiply when converting from a larger measurement unit to a smaller unit. For example, 6 yards converts to 18 feet when multiplied by the conversion factor 3 feet = 1 yard. When converting from a smaller unit to a larger one, you divide. For example, 13,200 feet converts to 2.5 miles when divided by the conversion factor 5280 feet = 1 mile.

Proportions can also be used to convert measurements. You can use the conversion chart given to you on your FCAT Reference Sheet. To make the conversion from one unit of measurement to another, use one conversion for one ratio.

**Example 1:** Jeff and his friends walked 3.5 kilometers in 1 day. How many meters did they walk?

Step 1: Set up a proportion using words: $\frac{\text{kilometers}}{\text{meters}} = \frac{\text{kilometers}}{\text{meters}}$.

Step 2: Use the conversion 1 kilometer = 1000 meters $\left(\frac{1 \text{ km}}{1000 \text{ m}}\right)$ as one ratio in your proportion.

Step 3: For the second ratio, use $\frac{3.5 \text{ km}}{\text{m}}$. Set the ratios equal to each other.

$$\frac{1 \text{ km}}{1000 \text{ m}} = \frac{3.5 \text{ km}}{\text{m}}.$$

Step 4: Cross-multiply $(1000 \times 3.5)$ and divide by 1.

Jeff and his friends walked 3500 m in 1 day.

## SCALE DRAWINGS

A drawing that is made to *scale* is proportional to the real object that it represents. Models of cars, airplanes, dollhouse furniture, and so on are made in proportion to the real things. Therefore, although one object is smaller than the other, the two look exactly the same.

**Example:** Assume two airplanes are proportional. Airplane $A$ is a scale model of airplane $B$. The length of airplane $A$ is 15 inches, and its width from wingtip to wingtip is 10 inches. Airplane $B$ is the actual airplane. Its length is 200 feet. What should be its width from wingtip to wingtip?

Since the information given tells us that the two airplanes are proportional, we know we can set up a proportion to solve for the missing width.

Step 1: Set up a proportion using words: $\frac{\text{length}}{\text{width}} = \frac{\text{length}}{\text{width}}$.

Step 2: Add the correct numbers: $\frac{15 \text{ in}}{10 \text{ in}} = \frac{200 \text{ ft}}{w \text{ ft}}$.

Step 3: Cross-multiply ($10 \times 200$), and divide by 15.

The wingspan of the actual airplane is a little over 133 ft.

## SAMPLE QUESTIONS

### Ratios, Rates, Proportions, and Scale Drawings

1. A jar held 85 black marbles and an unknown number of white marbles. The 20 marbles shown below were drawn at random from the jar and are in approximately the same proportion as the marbles in the original jar.

   Which is the most likely to be the number of *white marbles* that were originally in the jar?

   A. 100
   B. 150
   C. 180
   D. 200

2. Tammi bought a new fishbowl for her goldfish. The label on the bowl said that the bowl holds 64 ounces. How many quarts of water are in the fishbowl?

   F. 2
   G. 3
   H. 4
   I. 8

3. If the Haunted House is 0.4 km away, how many meters away is it?

   A. 4000
   B. 40
   C. 400
   D. 4

4. Jason made a model of the Eiffel Tower for the French Club. If the model was built so that 2 cm represents 7 meters, and Jason's model is 92 centimeters tall, how many meters tall is the actual Eiffel Tower?

5. The distance from Tampa to Port Charlotte is approximately 95 miles. If a Florida map has a scale of 1 inch = 17 miles, which of the following is closest to the distance between the two cities on the map?

   F. 5 in
   G. $5\frac{1}{8}$ in
   H. $5\frac{3}{8}$ in
   I. $5\frac{1}{2}$ in

6. Using the formula $d = rt$, fill in the entries missing from the table.

| Distance (miles) | Rate (miles per hour) | Time (hours) |
|---|---|---|
| 1. | 50 | 3 |
| 2. | 200 | 4 |
| 3. 210 | | 3.5 |
| 4. 500 | | 2.5 |
| 5. 261 | 58 | |
| 6. 187.5 | 75 | |

7. In her classroom, Mrs. Wilder has a scale model of one of the largest known blue whales. The scale used to build the model was 2 inches = 9 feet. If the scale model was 2 feet long, how long was the actual blue whale?

   A. 55 ft    C. 108 ft
   B. 99 ft    D. 220 ft

8. Shown below are the actual dimensions of a rectangular floor. Kellee uses a scale of 2 inches = 4 feet to make a scale drawing.

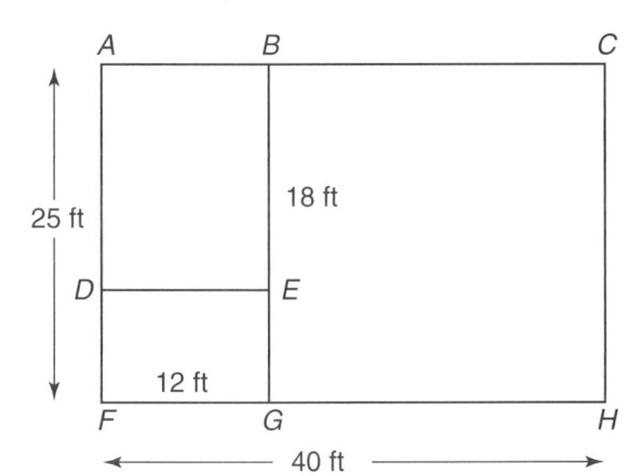


   What length, in inches, would $\overline{AD}$ be in the scale drawing?

9. If 8 hours out of 24 are spent sleeping, what decimal number represents the portion of the day that you are *awake*?

   F. 0.3    H. $0.\overline{3}$
   G. 0.6    I. $0.\overline{6}$

10. Danielle takes a photograph to a store to have the photograph enlarged. The original dimensions are 4 inches long by $2\frac{1}{2}$ inches wide. If the new photograph is 16 inches long, how wide, in inches, should it be?

11. Express the ratio $1\frac{1}{2}$ to 7 as a fraction in lowest terms.

12. Andrews Cement Company poured 250,000 pounds of cement during 2004. How many *tons* of cement did the company pour?

    A. 1250    C. 12.5
    B. 125    D. 1.25

13. The ratio of oil to vinegar in a salad dressing is 8:5. To the nearest ounce, how much oil must be blended with 7 ounces of vinegar for this recipe?

14. Bronze is an alloy of copper and tin. If a certain type of bronze requires 0.25 kilogram of copper per 1.0 kilogram of tin, how many kilograms of tin must be combined with 50 kilograms of copper?

15. If 3.5 grams of salt will dissolve in 10 liters of water, how many grams of salt will dissolve in 1 liter of water?

16. A flagpole casts a shadow 40 feet long. If a woman 5 feet tall casts a shadow 8 feet long at the same time and location, how tall is the flagpole?

F. 25 ft
G. 40 ft
H. 48 ft
I. 60 ft

17. The faces of Washington, Jefferson, Lincoln, and Theodore Roosevelt are carved into the side of Mount Rushmore in South Dakota. The ratio of the faces to the actual size of a man's face is 155:2. If Lincoln was 6 feet tall, how tall would the carving have to be in order to represent his entire body?

A. 12 ft
B. 215 ft
C. 310 ft
D. 465 ft

THINK
SOLVE
EXPLAIN

18. A laser printer prints 1 page every 3 seconds.

Part A: How many pages can it print in 1 minute? Show your work, or explain how you arrived at your answer.

________________________________

________________________________

________________________________

Part B: How many pages can it print in three quarters of an hour? Show your work, or explain how you arrived at your answer.

________________________________

________________________________

________________________________

THINK
SOLVE
EXPLAIN

19. Dylan is swimming in the olympic-size pool shown below.

50 m

25 m

How many laps will he have to swim if he wants to swim 2 kilometers? Show your work, or explain how you arrived at your answer.

________________________________

________________________________

________________________________

THINK
SOLVE
EXPLAIN

20. A bottle-nosed dolphin travels 15 miles off the coast of Pine Island at an average speed of 35 miles per hour (mph). How far will the dolphin travel in 24 minutes?

________________________________

________________________________

________________________________

## ANSWERS TO SAMPLE QUESTIONS

1. **D.** Set up a proportion:

$$\begin{array}{ccc} \text{Jar} & & \text{Sample} \\ \frac{85 \text{ black}}{\text{white}} & = & \frac{6 \text{ black}}{14 \text{ white}} \end{array}$$

$$85 \times 14 = 6w$$
$$1190 = 6w$$
$$1190 \div 6 = 198$$

198 is closest to 200.

2. **F.** You are converting from a smaller unit (ounces) to a larger unit (quarts); 32 oz = 1 qt. Divide 64 oz by 32.

3. **C.** Since 1 km = 1000 m:

$\frac{1 \text{ km}}{1000 \text{ m}} = \frac{0.4 \text{ km}}{x \text{ m}}$;

$(1000 \times 0.4 \div 1) = 400$ m.
An alternative method is to multiply 0.4 by 1000 because you are converting from a larger unit (kilometers) to a smaller unit (meters).

4. **322.** Set up a proportion. Use the scale as one ratio, and the actual measurements of the model and the real Eiffel Tower as the second ratio.

$\frac{7 \text{ m}}{2 \text{ cm}} = \frac{x \text{ m}}{92 \text{ cm}}$.

Solve by cross-multiplying and dividing by 2: $(7 \times 92) \div 2 = 322$.

5. **I.** Use the map scale to set up a proportion:

$\frac{1 \text{ in}}{17 \text{ mi}} = \frac{x \text{ in}}{95 \text{ mi}}$;

$(95 \times 1) \div 17 = 5\frac{1}{2}$.

6.

| Distance (miles) | Rate (miles per hour) | Time (hours) |
|---|---|---|
| 1. **150** | 50 | 3 |
| 2. **800** | 200 | 4 |
| 3. 210 | **60** | 3.5 |
| 4. 500 | **200** | 2.5 |
| 5. 261 | 58 | **4.5** |
| 6. 187.5 | 75 | **2.5** |

7. **C.** Use the scale to set up a proportion: $\frac{2 \text{ in}}{9 \text{ ft}} = \frac{24 \text{ in}}{x \text{ ft}}$.
Notice that the 2-ft label on the whale was changed to 24 inches in the proportion so that the ratios are in equal units. Multiply diagonally and divide by the remaining number:
$9 \times 24 \div 2 = 216 \div 2 = 108$.

8. **9.** Use the scale to set up a proportion: $\frac{2 \text{ in}}{4 \text{ ft}} = \frac{x \text{ in}}{18 \text{ ft}}$;
$(18 \times 2) \div 4 = 9$.

9. **I.** You are awake 16 hr out of 24.
The fraction $\frac{16}{24} = 0.\overline{6}$.

10. **10.** Use the proportion:
$\frac{4 \text{ long}}{2.5 \text{ wide}} = \frac{16 \text{ long}}{x \text{ wide}}$;
$(2.5 \times 16) \div 4 = 10$ in.

11. $\frac{\mathbf{3}}{\mathbf{14}}$. The original fraction is $\frac{1\frac{1}{2}}{7}$. Multiply the numerator and denominator by 2 to eliminate the fraction in the numerator.

12. **B.** You are converting from a smaller to a larger unit. Since 1 ton = 2000 pounds, divide 250,000 by 2000. Alternatively, you can use the proportion $\frac{1 \text{ ton}}{2000 \text{ lb}} = \frac{x \text{ tons}}{250{,}000 \text{ lb}}$; $(250{,}000 \times 1) \div 2000 = 125$.

13. **11.** Use the proportion: $\frac{8 \text{ oil}}{5 \text{ vinegar}} = \frac{x \text{ oil}}{7 \text{ vinegar}}$; $(7 \times 8) \div 5 = 11.2$, which rounds to 11 oz.

14. **200 kg.** Use the proportion: $\frac{0.25 \text{ copper}}{1 \text{ tin}} = \frac{50 \text{ copper}}{x \text{ tin}}$; $(1 \times 50) \div 0.25$.

15. **0.35.** Use the proportion: $\frac{3.5 \text{ salt}}{10 \text{ water}} = \frac{x \text{ salt}}{1 \text{ water}}$; $(1 \times 3.5) \div 10 = 0.35$.

16. **F.** Use the proportion: $\frac{5}{8} = \frac{x}{40}$; $(40 \times 5) \div 8 = 25$.

17. **D.** Use the proportion: $\frac{155 \text{ carving}}{2 \text{ actual}} = \frac{x \text{ carving}}{6 \text{ actual}}$; $x = 465$.

18. Part A. **20.** Set up a proportion. The information is given in seconds, but the question asks for an answer in minutes: $\frac{1 \text{ page}}{3 \text{ sec}} = \frac{x \text{ pages}}{60 \text{ sec}}$. Since 1 minute = 60 sec, you can use a proportion to change 1 page in 3 seconds to 20 pages in 60 sec: $\frac{1 \text{ page}}{3 \text{ sec}} = \frac{20 \text{ pages}}{60 \text{ sec}}$.
Part B. **900.** The printer will print 20 pages in 1 min and $20 \times 45$ or 900 pages in 45 min (three quarters of an hour = 45 min).

19. **40.** This problem requires two major steps. In the first step, you need to change kilometers to meters in order to answer the question. Since 1 km = 1000 m, 2 km = 2000 m. In the second step, use the information from Step 1 to set up a proportion. If 1 lap = 50 m, then 40 laps = 2000 m: $\frac{1 \text{ lap}}{50 \text{ m}} = \frac{x \text{ laps}}{2000 \text{ m}}$; $x = 40$ laps.

20. **14 mi.** You do not need to know how far off the coast the dolphin is traveling. You can use a proportion to solve the problem: $\frac{35 \text{ mi}}{60 \text{ min}} = \frac{x \text{ mi}}{24 \text{ min}}$; $x = 14$ mi. Note that 35 mph has been converted to 35 mi in 60 min in the first ratio.

# Lesson 5

# Percents

## EQUIVALENT PERCENTS, DECIMALS, AND FRACTIONS

Write 12.5% in decimal form.

To change a percent to a decimal, remove the percent sign and move the decimal point exactly two places to the *left*. For example, 12.5% = 0.125.

Percents are related to decimals and fractions. You should be able to write percents as decimals **and** as fractions.

To change a decimal to a percent, reverse the process. Move the decimal point exactly two places to the *right* and add a percent sign.

**Examples:** 0.998 = 99.8%

4.18 = 418%

To change a percent to a fraction, write the number over 100, drop the percent sign, and put the fraction into lowest terms.

**Example 1:** Write 45% as a fraction.

$$45\% = \frac{45}{100} = \frac{9}{20}$$

**Example 2:** Write 17.7% as a fraction.

$17.7\% = \frac{17.7}{100}$, which is correctly expressed as $\frac{177}{1000}$.

(Multiply the numerator and the denominator by 10 so that you will not have a decimal point in the fraction.)

## USING PROPORTIONS TO SOLVE PERCENT PROBLEMS

An excellent method for changing a fraction to a percent or a percent to a fraction is to use a proportion. The proportion should look like this: $\frac{\text{Part}}{\text{Whole}} = \frac{\%}{100}$.

**Example 1:** Change $\frac{2}{3}$ to a percent and then to a decimal.

Use $\frac{\text{Part}}{\text{Whole}} = \frac{\%}{100}$. Fill in the numbers: $\frac{2}{3} = \frac{\%}{100}$. Multiply $(100 \times 2)$, and divide by 3.

The answer in your calculator's display will be $66.\overline{6}$, which represents the percent. Change this to a decimal by moving the decimal point two places to the left: $0.66\overline{6}$, which should be written as $0.\overline{6}$.

**Example 2:** Angie's gas tank holds 20 gallons. If she has used 12 gallons, what percent is left?

Use $\frac{\text{Part}}{\text{Whole}} = \frac{\%}{100}$. Angie has 8 gallons remaining out of 20: $\frac{8}{20} = \frac{\%}{100}$. Multiply $(100 \times 8)$ and divide by 20. She has 40% remaining.

**Example 3:** Jose went to the new mall and spent 75% of his money. If he bought $150 worth of merchandise, how much money did he start with?

Use $\frac{\text{Part}}{\text{Whole}} = \frac{\%}{100}$. Jose spent part of his money, or $150. You are given the percent in this problem, so substitute 75% for % in the proportion:
$\frac{150}{\text{Whole}} = \frac{75\%}{100}$.
Multiply $(100 \times 150)$ and divide by 75. Jose started with $200.

## Percent of Increase and Percent of Decrease

To calculate a percent of increase or decrease, you can use a proportion similar to the one above:

$$\frac{\text{Amount of change}}{\text{Original amount}} = \frac{\%}{100}.$$

**Example 1:** A baby weighed 7.4 pounds at birth and 9.0 pounds at the age of 1 month. By what percent did the baby's weight increase? To find the amount of change, subtract.

$$\frac{\text{Amount of change}}{\text{Original amount}} = \frac{\%}{100} = \frac{9.0-7.4}{7.4} = \frac{\%}{100} = \frac{1.6}{7.4} = \frac{\%}{100} = 21.6\%$$

**Example 2:** Mr. Pattison bought LAS stock at \$45 a share and sold it at \$30. By what percent did the stock price decrease?

$$\frac{\text{Amount of change}}{\text{Original amount}} = \frac{\%}{100} = \frac{45-30}{45} = \frac{\%}{100} = \frac{15}{45} = \frac{\%}{100} = 33\frac{1}{3}\%$$

## The FCAT Calculator Percent Key

The FCAT calculator has a percentage key, [ % ]. You can use this key to find (multiply) percents, or you can change the percent to a decimal and then multiply.

**Example 1:** Florida charges 7% sales tax. How much tax will there be on a purchase of \$525?

Method 1: [5] [2] [5] [×] [7] [%] (don't hit =). \$36.75

Method 2: [5] [2] [5] [×] [.] [0] [7] [=]. \$36.75

**Example 2:** Maria bought new tires for her car. She paid \$265 for four tires plus 7% tax. What was her total bill?

Method 1: [2] [6] [5] [+] [7] [%] (don't hit =). \$283.55

Method 2: [2] [6] [5] [×] [1] [.] [0] [7] [=]. \$283.55

**Example 3:** Patrick bought a new car. The sticker price was $24,500, but the dealer gave him a 12% discount. What was the price of the new car?

Method 1: [2] [4] [5] [0] [0] [−] [1] [2] [%] (don't hit =). $21,560

Method 2: [2] [4] [5] [0] [0] [×] [.] [8] [8] [=]. $21,560

Method 3: [2] [4] [5] [0] [0] [×] [.] [1] [2] [=]
[M+] [2] [4] [5] [0] [0] [−] [MRC] [=]. $21,560

Method 3 makes use of the memory in the FCAT calculator. The key [M+] (memory add) allows you to add your answer to the memory and hold it while you make other calculations. The key [MRC] (memory recall) allows you to retrieve this number when you need it later.

# SAMPLE QUESTIONS

## Percents

1. Fill in the blanks in the table with equivalent fractions, decimals, or percents.

| Fraction | Decimal | Percent |
|---|---|---|
| $\frac{3}{5}$ | | 60 |
| | $.\overline{3}$ | |
| | | 25 |
| $\frac{3}{100}$ | | |
| $\frac{7}{20}$ | | 3.5 |
| | 2.5 | |
| | 0.125 | |

2. An environmentalist has discovered an 11% increase in the number of alligators in a section of the Florida Everglades during the past 3 years. If there were 112,500 alligators 3 years ago, what is the approximate number of alligators in this area now?

   A. 1330
   B. 95,625
   C. 122,511
   D. 125,000

3. The asking price for a house was $115,000 before a 15% discount was offered. What is the new price of the house?

4. Students in Mrs. Rockwell's afternoon class are completing math projects. Jorge has finished 55% of his project. Maryann has completed 0.65 of her project. Lucy has completed $\frac{3}{5}$ and Justin has completed $\frac{2}{3}$ of their projects. Which list shows the students in order from the student who has completed the most to the one who has completed the least?

   F. Maryann, Jorge, Lucy, Justin
   G. Jorge, Maryann, Justin, Lucy
   H. Justin, Maryann, Lucy, Jorge
   I. Justin, Lucy, Jorge, Maryann

5. Laurie's group is conducting an experiment to see whose hair sample will grow the most bacteria in a Petri dish. At the end of 1 week, the results were as follows:

   Laurie's bacteria covered 0.35 of her dish.

   Sam's bacteria covered $\frac{3}{8}$ of his dish.

   Nancy's bacteria covered $\frac{2}{5}$ of her dish.

   Maurio's bacteria covered 37% of his dish.

   Whose dish had grown the *fewest* bacteria?

   A. Laurie's
   B. Sam's
   C. Nancy's
   D. Maurio's

6. Erica was assigned to reprice items for her aunt's online gift-shop sale. Some of the items are shown in the table.

| Item on Sale | Original Price | Percent Discount | Sale Price | Percent Sales Tax | Total Amount |
|---|---|---|---|---|---|
| Crystal lamp | \$150 | 10% | | 6% | |
| Ceramic vase | \$45 | 25% | | 4% | |
| Deco cat | \$110 | 20% | | 8% | |
| Jewelry chest | \$59.95 | 15% | | 7% | |
| Piggy bank | \$35.50 | 30% | | 5% | |
| Wooden elephant | \$99 | 33⅓% | | 5½% | |
| Crystal frame | \$75 | 40% | | 4½% | |
| Painting | \$125 | 10% | | 7% | |

For each item, fill in the sale price and the total amount after the sales tax is added. Round your answers to the nearest cent.

7. A local bank loaned Lou \$500 for 1 year at 12% interest per year. What was the total amount Lou had to repay at the end of the year?

8. If Lou decided to pay off the loan in Problem 7 after 4 months, how much interest would he owe?

9. Gold ore mined from the Ankara Gold Mine is 0.85% pure. How many grams of gold are there in 10,000 grams of ore?

   F. 8.5 H. 85
   G. 850 I. 8500

10. Don and Cheryl bought their house 13 years ago. When they sold it, they received 210% of the original purchase price. If the original price was \$70,000, how much money did they receive for the house?

    A. \$70,210 C. \$120,000
    B. \$147,000 D. \$210,000

11. A store in West Palm Beach has a clearance sale offering 55% off every item. Which of the following is NOT another way to write 55%?

    F. $\frac{55}{100}$ H. $\frac{11}{20}$
    G. $\frac{5}{11}$ I. 0.55

12. Mr. Gonzalez took his family of five to a soccer game. Each ticket cost $12.00. Mr. Gonzalez bought each person a hot dog for $3.00 and a drink for $2.50. He was charged 7% tax on everything he bought. What was the entire cost for the trip? Round your answer to the nearest penny.

13. Marcel's cell phone bill was $39.99 last month. This month the bill was $44.79. By what percent did Marcel's bill increase this month?

    A. 12 C. 14
    B. 13 D. 15

14. Which circle graph contains a white portion of approximately 60%?

F. 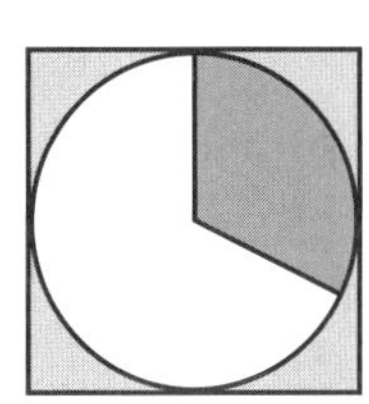

H. 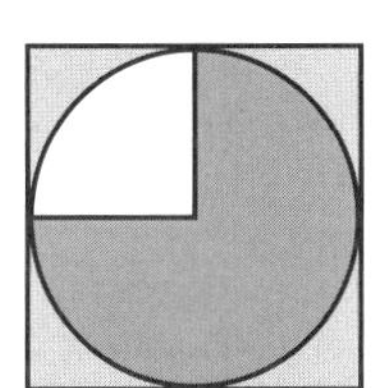

G. 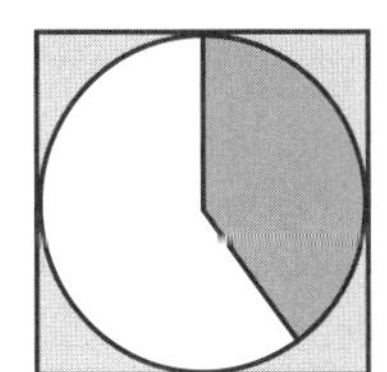

I. 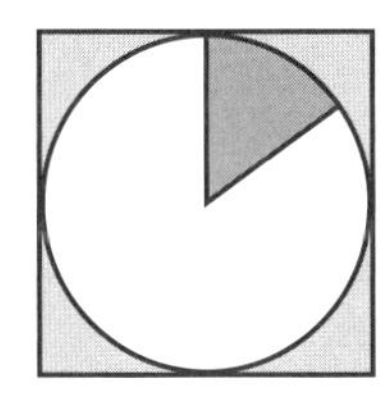

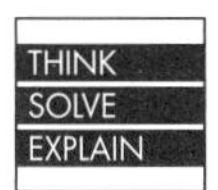


15. Takisha wanted to purchase the following items:

| Item | Price | Tax |
|---|---|---|
| Gold necklace | $99.99 | 7% |
| Sapphire ring | $149.59 | 7% |

She raised enough money to make these purchases in 5 weeks. If she charged $4.00 an hour for babysitting, ESTIMATE how many hours she had to babysit per week to raise the money. Explain your answer in words, or show your work.

______________________________

______________________________

______________________________

16. Men outnumber women in 4 out of 50 states. In what percent of the states do women outnumber men?

17. The price of an item before tax was $45. The sales tax was $3.15. What percent of tax was charged?

18. The price of a milk shake at Mr. Shake increased from $1.95 to $2.34. What was the percent of increase?

    A. 15% C. 20%
    B. 25% D. 35%

19. During a sale, a 10-speed bike was marked down from $150 to $90. What was the percent of discount?

    F. 60%  H. 40%
    G. 50%  I. 30%

20. A math experiment involved a paper bag with 8 orange tiles, 5 yellow tiles, 3 blue tiles, and 2 red tiles. Approximately what percent of the cubes were red?

    A. 2  C. 22
    B. 11  D. 40

## ANSWERS TO SAMPLE QUESTIONS

1.

| Fraction | Decimal | Percent |
|---|---|---|
| $\frac{3}{5}$ | 0.6 | 60 |
| $\frac{1}{3}$ | $.\overline{3}$ | $33\frac{1}{3}$ |
| $\frac{1}{4}$ | 0.25 | 25 |
| $\frac{3}{100}$ | 0.03 | 3 |
| $\frac{7}{20}$ | 0.035 | 3.5 |
| $\frac{250}{100}$ | 2.5 | 250 |
| $\frac{1}{8}$ | 0.125 | 12.5 |

2. D is closest.
   112,500 + [1] [1] [%] = 124,875.

3. **97,750.**
   $115,000 – [1] [5] [%] = $97,750.

4. **H.** Put each number into decimal form and compare: Justin = $\frac{2}{3}$ = 0.67; Maryann = 0.65; Lucy = $\frac{3}{5}$ = 0.6; Jorge = 55% = 0.55.

5. **A.** Put each number into decimal form and compare. Laurie = 0.35; Sam = $\frac{3}{8}$ = 0.375; Nancy = $\frac{2}{5}$ = 0.4; and Maurio = 37% = 0.37.

6.

| Item on Sale | Original Price | Percent Discount | Sale Price | Percent Sales Tax | Total Amount |
|---|---|---|---|---|---|
| Crystal lamp | $150 | 10% | $135 | 6% | $143.10 |
| Ceramic vase | $45 | 25% | $33.75 | 4% | $35.10 |
| Deco cat | $110 | 20% | $88.00 | 8% | $95.04 |
| Jewelry chest | $59.95 | 15% | $50.96 | 7% | $54.52 |
| Piggy bank | $35.50 | 30% | $24.85 | 5% | $26.09 |
| Wooden elephant | $99 | 33⅓% | $66.00 | 5½% | $69.63 |
| Crystal frame | $75 | 40% | $45 | 4½% | $47.02 |
| Painting | $125 | 10% | $112.50 | 7% | $120.38 |

7. **560.** 500 + [1] [2] [%] = \$560.

8. **20.** 500 × [1] [2] [%] = \$60.
Divide by 12 for interest for 1 month, and multiply by 4 for interest for 4 months.

9. **H.**
10,000 × [.] [8] [5] [%] = 85.

10. **B.** 70,000 × [2] [1] [0] [%] = \$147,000.

11. **G.** The amount off, 55%, can be written as a fraction, $\frac{55}{100}$, which is $\frac{11}{20}$ in lowest terms, and as 0.55 as a decimal.

12. **93.63.** Multiply the costs by 5:
5(12.00 + 3.00 + 2.50) =
5(17.50 = \$87.50. Add 7%:
87.50 + 7% = \$93.63.

13. **A.** Use the proportion
$\frac{\text{Amount of change}}{\text{Original amount}} = \frac{\%}{100} =$
$\frac{\$4.80}{\$39.99} = \frac{\%}{100} = 12\%.$

14. **G.** Sixty percent is slightly more than half.

15. Takisha worked about **13.5** hr per week. She needed to raise about \$250 + 7% tax or about \$270. Divide \$270 by 5 to get the amount she needed to earn per week (\$54). Divide \$54 by \$4 per hour to get the number of hours.

16. **92.** Men outnumber women in 4 out of 50 states. Written as a fraction, $\frac{4}{50} = \frac{8}{100} = 8\%$.
Men outnumber women in 8% of the states. Therefore, women outnumber men in the other 92%.

17. **7.** $\frac{3.15}{45} = \frac{\%}{100} = 7.0\%$

18. **C.** $\frac{\text{Amount of change}}{\text{Original amount}} = \frac{\%}{100} =$
$\frac{0.39}{1.95} = \frac{\%}{100} = 20\%$

19. **H.** $\frac{\text{Amount of change}}{\text{Original amount}} = \frac{\%}{100} =$
$\frac{60}{150} = \frac{\%}{100} = 40\%$

20. **G.** $\frac{\text{Part}}{\text{Whole}} = \frac{\%}{100} =$
$\frac{2}{18} = \frac{\%}{100} = 11\%$

# Lesson 6

# Geometry Basics

## POINT

A *point* is the most basic unit of geometry. It is represented by a dot and named by a capital letter, as shown. A point has no dimensions: no thickness, no width, no length.

## LINE

A straight *line* is endless in length. It is usually named using two points on the line or by a single letter in italics, as shown. Notice, to the right of the first line, the double-pointed arrow shown over *AB* in the name. A lines has one dimension: length.

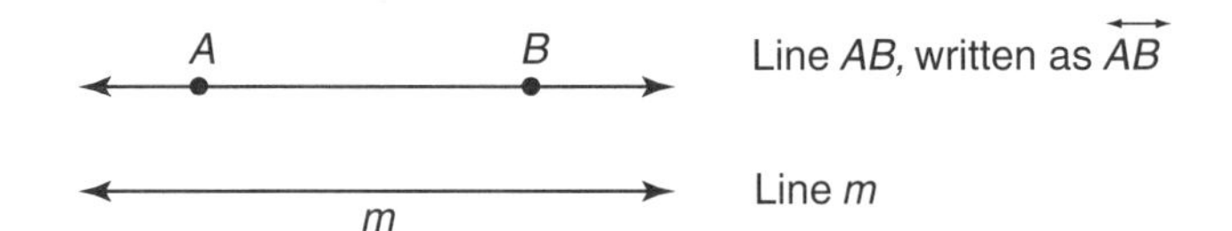


## PLANE

A *plane* is a two-dimensional (flat) geometric surface. It has endless length and width. It is represented by a four-sided figure like the ones shown and is named by one capital letter or by three capital letters. A plane has two dimensions: length and width.

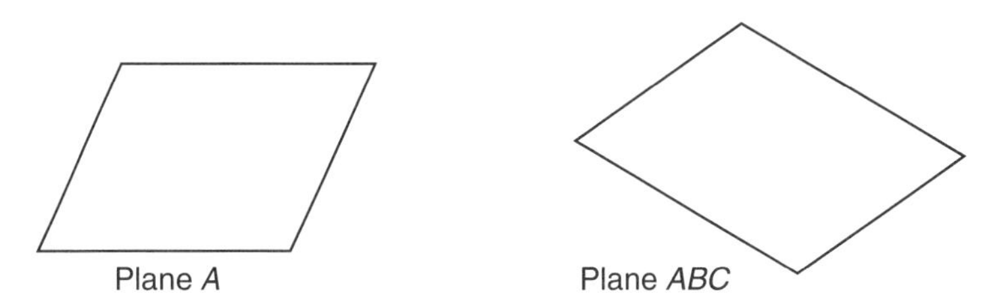

## LINE SEGMENT

A *line segment* is a piece of a line. It has a beginning and an end and is named by the points that represent the beginning and the end. The line segment shown is line segment *CD* and is named $\overline{CD}$.

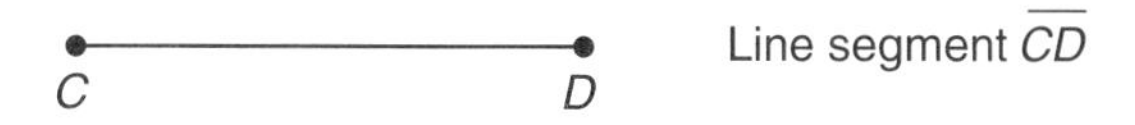


Line segment $\overline{CD}$

## RAY

A *ray* is also a piece of a line. It has a beginning (called an endpoint) and continues on endlessly in one direction. It is named by two letters, the first letter for the ray's beginning and the second letter for another point on the ray, as shown. Pay close attention to how a ray is named, as naming can be a little tricky. Remember: the nonarrow part of the ray symbol is located over the endpoint letter.

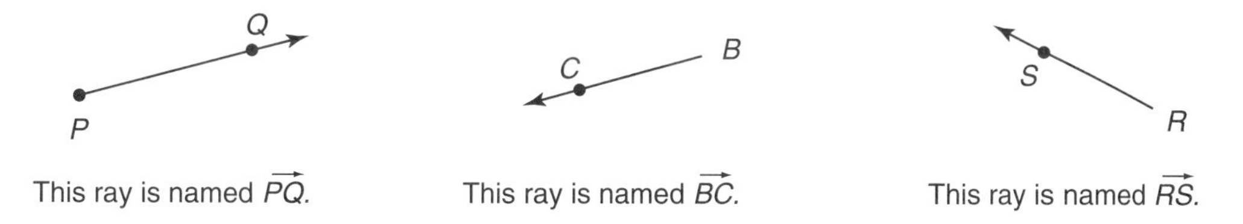


This ray is named $\overrightarrow{PQ}$. This ray is named $\overrightarrow{BC}$. This ray is named $\overrightarrow{RS}$.

## MIDPOINT

A *midpoint* of a line segment is the point that is exactly in the middle between the two endpoints, as shown. Point *M* represents the midpoint of line segment $\overline{PO}$.

## THREE TYPES OF LINES

### Parallel Lines

Lines that lie in the same plane and will never intersect are called *parallel lines*. Line $\ell$ and line $m$ are parallel lines. Parallel lines can be marked with little arrowheads, as shown, to indicate that they are parallel. To write that line $\ell$ is parallel to line $m$, write $\ell \parallel m$.

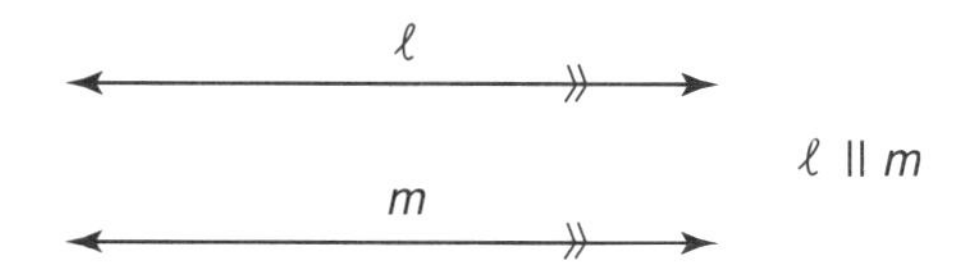

## Perpendicular Lines

When lines, line segments, or rays intersect at a 90° angle (a right angle), they are *perpendicular*. The intersection (the point where they cross) is usually marked by a small box. In the diagram, line $AB$ is perpendicular ($\perp$) to ray $CD$.

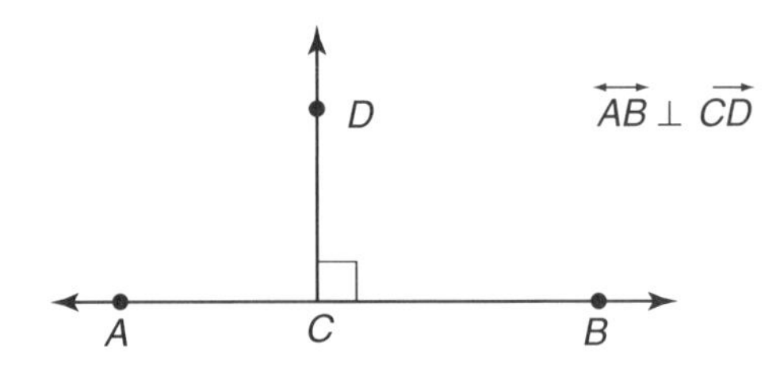


## Skew Lines

Lines that lie in different planes and do not intersect, as shown, are called *skew lines*.

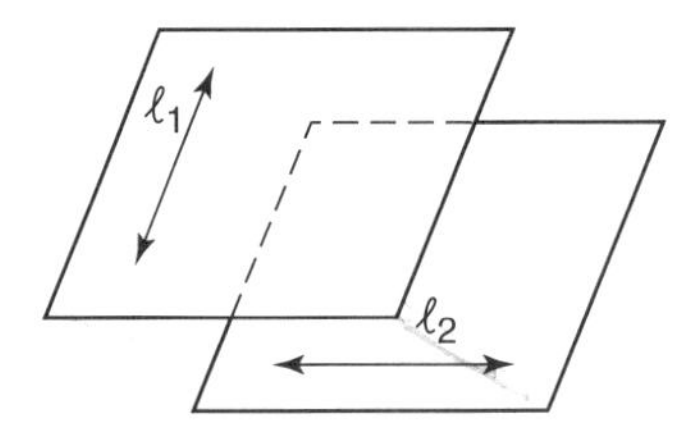


We say that line 1 ($\ell_1$) and line 2 ($\ell_2$) are skew.

In the accompanying drawing, $\overline{AB}$ and $\overline{DF}$ are skew lines because they lie in different planes and do not intersect. $\overline{AB}$ and $\overline{BD}$ are perpendicular lines, while $\overline{EF}$ and $\overline{GH}$ are parallel lines.

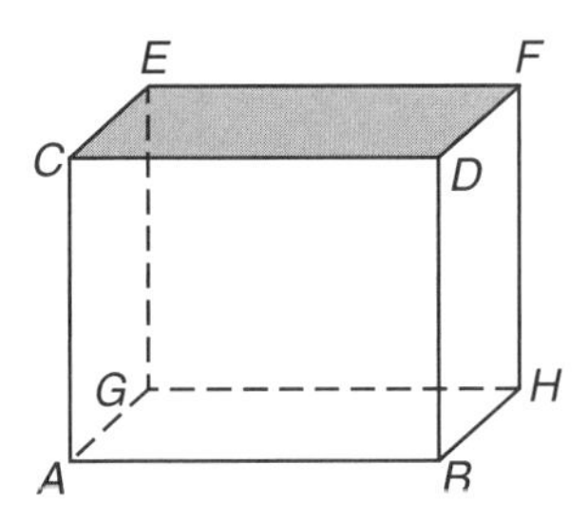

# SAMPLE QUESTIONS

## Geometry Basics

1. Which of the following does NOT represent the correct name for the drawing shown?

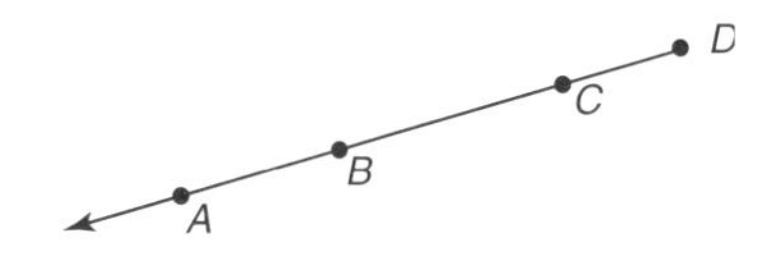


A. $\overrightarrow{DC}$ C. $\overrightarrow{DA}$

B. $\overrightarrow{DB}$ D. $\overrightarrow{AD}$

2. Select the diagram that represents $\overline{QR}$.

F. Q R

G. Q R

H. Q R

I. Q R

3. The statement $\overleftrightarrow{RP} \perp \overline{MN}$ is illustrated by which of the following drawings?

A.

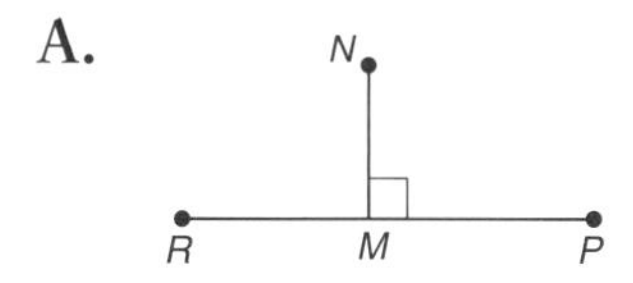


B.

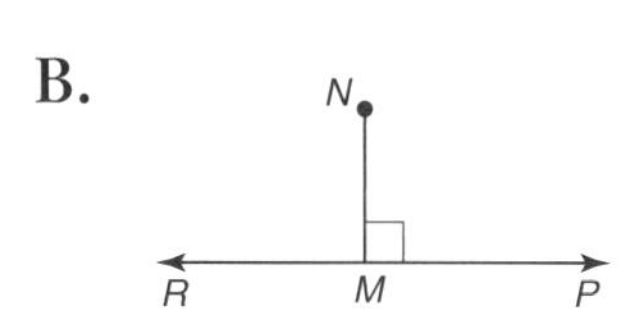


C.

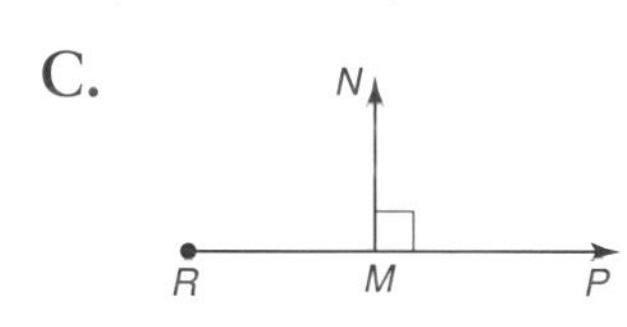


D.

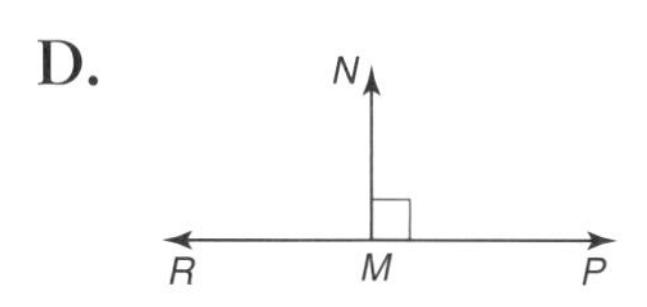


4. Select the geometry notation that best matches the drawing shown.

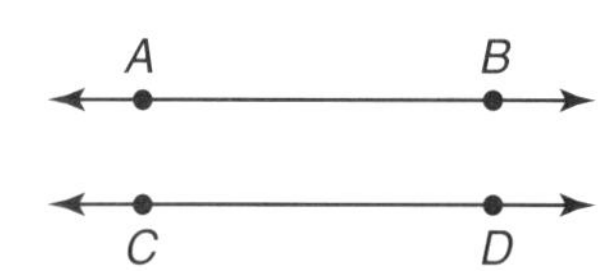


F. $\overleftrightarrow{AB} \perp \overleftrightarrow{CD}$

G. $\overline{AB} \parallel \overline{CD}$

H. $\overleftrightarrow{AB} \parallel \overleftrightarrow{CD}$

I. $\overline{AB} \perp \overline{CD}$

5. The most basic unit of geometry is a

A. line C. plane
B. point D. ray

6. In the drawing, point $M$ represents the midpoint of line segment $\overline{AB}$. Which two line segments are of equal length?

F. $\overline{AB}$ and $\overline{MC}$
G. $\overline{BC}$ and $\overline{MB}$
H. $\overline{MC}$ and $\overline{AM}$
I. $\overline{AM}$ and $\overline{BC}$

7. **M A H K**

Which of the letters shown above contains both parallel and perpendicular line segments?

A. **K** C. **A**
B. **H** D. **M**

8. In the rectangular prism shown, $\overline{CG}$ has what relationship to $\overline{DC}$?

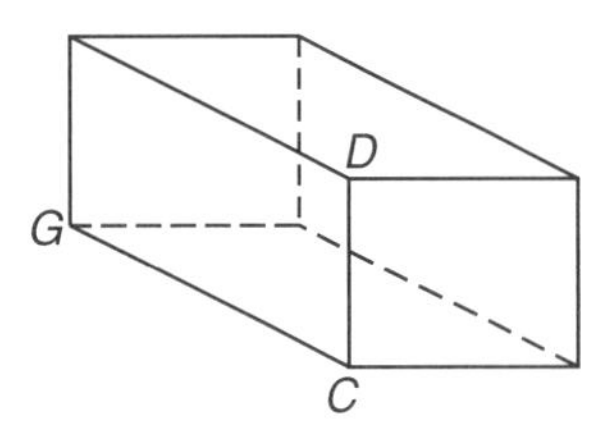


F. parallel
G. perpendicular
H. skew
I. obtuse

9. What is the coordinate of the midpoint of $\overline{AB}$?

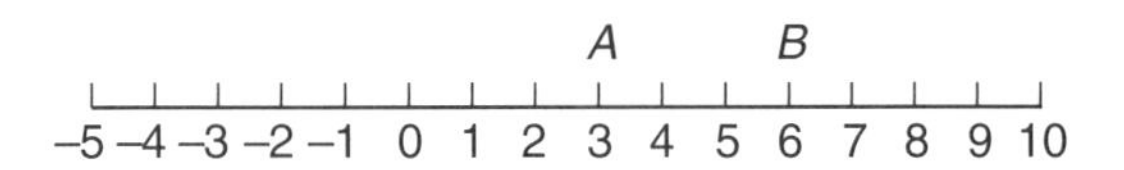


10. If the endpoints of a line segment are located at –4 and 8, where is the midpoint located?

## ANSWERS TO SAMPLE QUESTIONS

1. **D.** A ray is named by two letters—endpoint first and then another point on the line.

2. F. A line segment is represented by two letters with a short line segment above.

3. **B.** $\overleftrightarrow{RP}$ represents a line, and $\overline{MN}$ represents a line segment.

4. **H.** The drawing shows two parallel (||) lines.

5. **B.** The most basic unit of geometry is a point.

6. **H.** The midpoint divides a line segment into two equal parts. From the information given, only $\overline{AM}$ and $\overline{MC}$ are equal.

7. **B.** The letter **H** contains both parallel and perpendicular line segments.

8. **G.** The two line segments intersect at a 90-degree angle.

9. **4.5.** The midpoint of a line segment is found by adding the coordinates of the endpoints and dividing by 2: $\frac{3+6}{2} = \frac{9}{2} = 4.5$.

10. **2.** The midpoint of a line segment is found by adding the coordinates of the endpoints and dividing by 2: $\frac{-4+8}{2} = \frac{4}{2} = 2$.

# Lesson 7

# Angles

An *angle* is formed by two rays that share the same endpoint, as shown. The shared endpoint is called the *vertex*, and the rays are the *sides* of the angle.

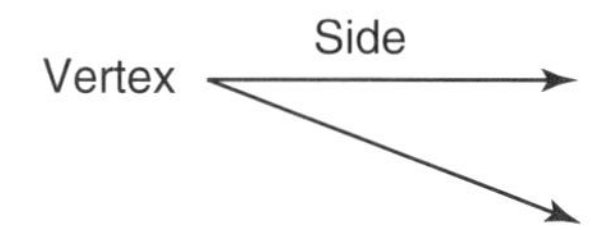


The unit of measurement for angles is degrees. As shown in the diagrams, an angle can be named using a single number as in $\angle 1$ (read as "angle one"), a single letter (which represents the vertex), as in $\angle B$, or three letters, as in $\angle ABC$. If an angle is named by three letters, the center letter, in this case $B$, represents the vertex of the angle.

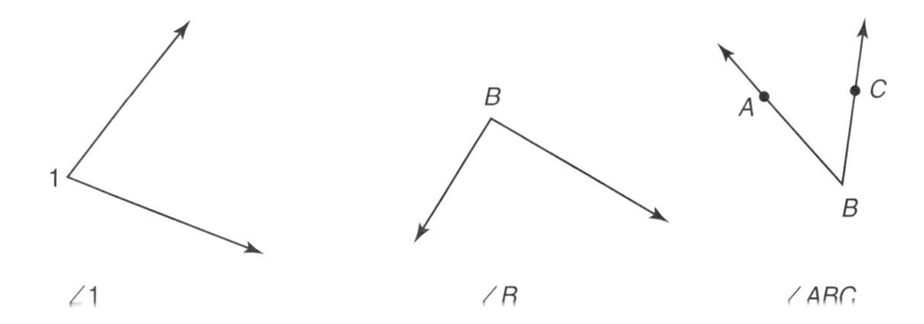


In the diagram, $\angle 2$ represents the same angle as $\angle ABC$. You should not use the name $\angle B$ for this figure because $\angle B$ could mean $\angle 1$, $\angle 2$, $\angle 3$, or $\angle 4$. Whenever you see a complex diagram like this one, it is a good idea to name an angle using all three letters to avoid confusion.

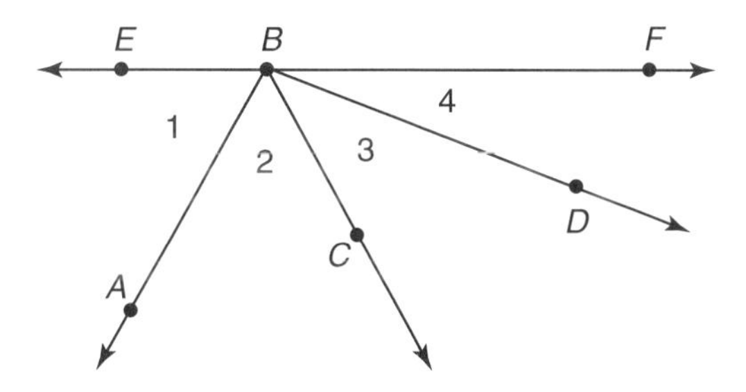

## KINDS OF ANGLES

### Right Angle

A *right angle* has a measure of 90°. In the diagram, the small block inside the angle at the vertex tells you that this is a right angle. Unless you see a small block or you are told that an angle is a right angle, do not assume from a drawing that the angle measures 90°. For this diagram, we say that the measure of ∠1 equals 90°. This is written as m∠1 = 90°.

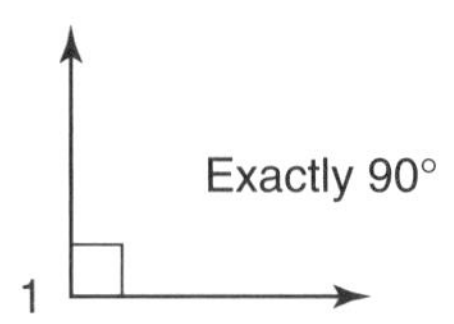


### Obtuse Angle

An *obtuse angle* measures more than 90° but less than 180°.

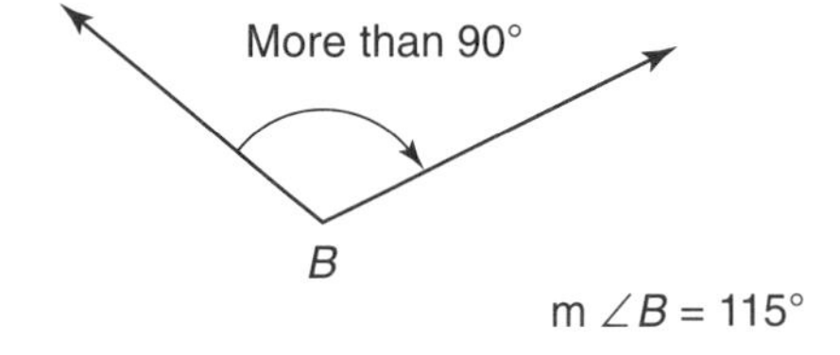


### Acute Angle

An *acute angle* measures less than 90°.

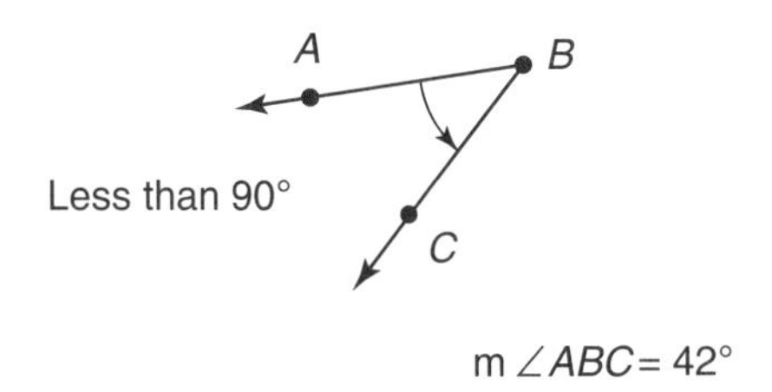


### Straight Angle

A *straight angle* measures exactly 180°. It looks like a line, but is labeled with three points. In the diagram, ∠*PQR* is a straight angle.

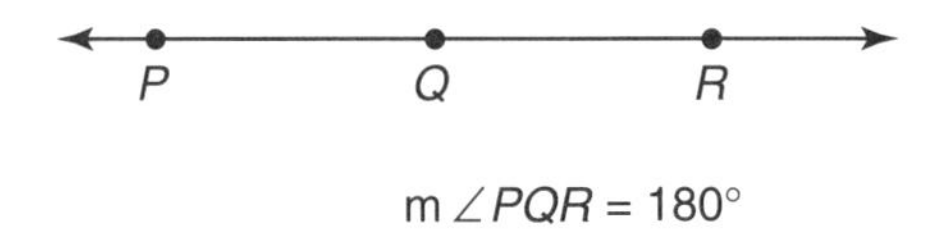

## Reflex Angle

A *reflex angle* measures more than 180° but less than 360°.

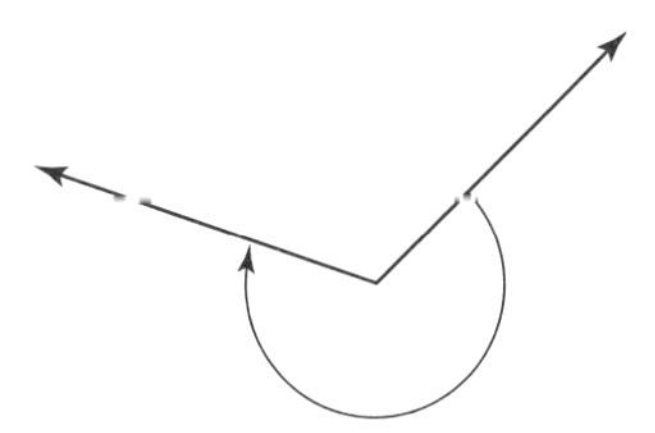

## Adjacent Angles

*Adjacent angles* share both a vertex and a side. The angles do not overlap. In the diagram, ∠1 and ∠2 are adjacent angles because they share vertex $B$ and side $\overline{BD}$.

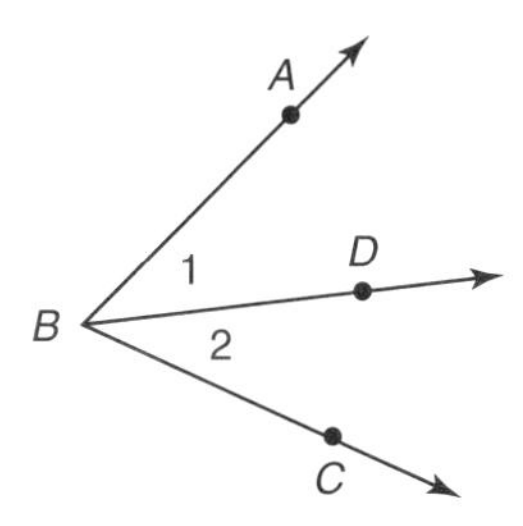


## Congruent Angles

*Congruent angles* have the same measure. In the diagram, the symbol ≅ indicates that angles $A$ and $B$ have the same measure and are congruent.

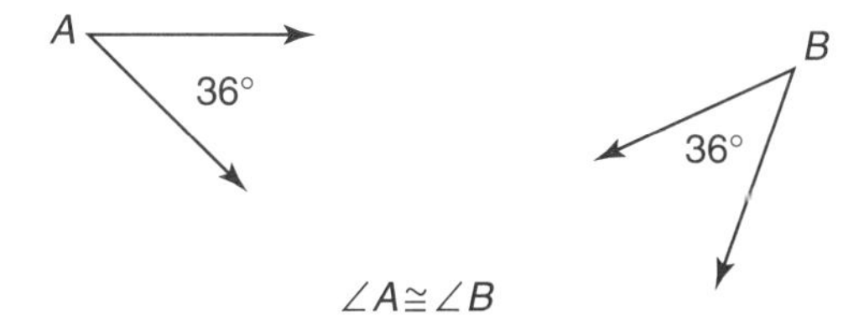


## Complementary Angles

*Complementary angles* are two angles whose measures add to 90°. The diagram indicates that m∠1 + m∠2 = 90° because ∠$DEF$ is a right angle. Therefore, ∠1 and ∠2 are complementary.

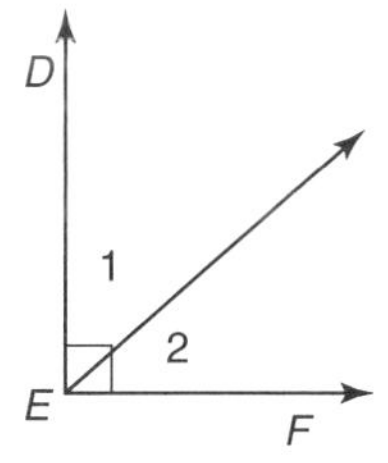


m∠1 + m∠2 = 90°

Complementary angles do not have to be adjacent. Because the measures of angles *A* and *B* in this diagram add to 90°, the angles are complementary.

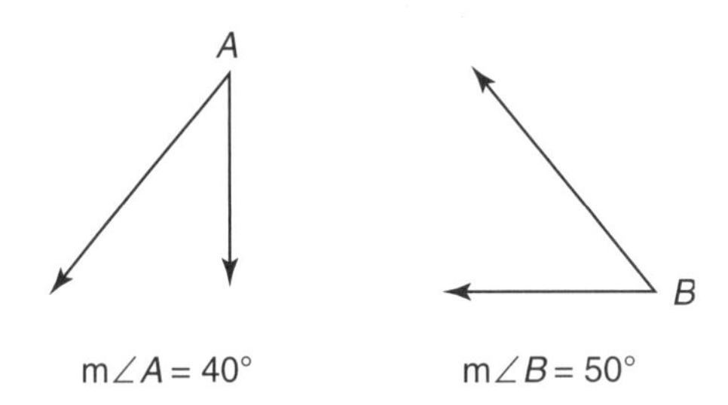


### Supplementary Angles

*Supplementary angles* are two angles whose measures add to 180°. The diagram indicates that m∠1 + m∠2 = 180° because the two angles form a linear pair. Therefore, ∠1 and ∠2 are supplementary.

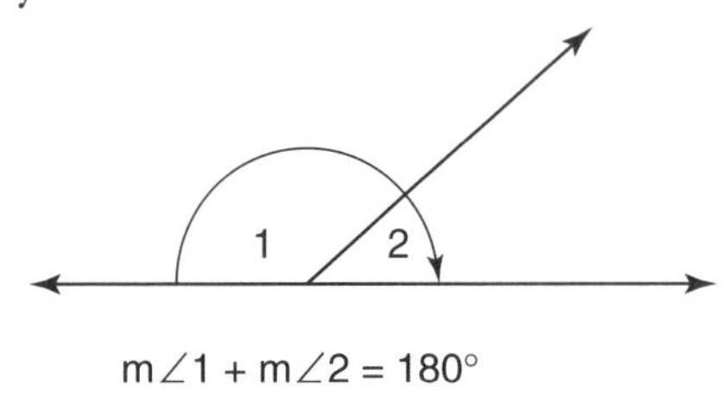


m∠1 + m∠2 = 180°

Supplementary angles do not have to be adjacent. Because the measures of angles *A* and *B* in this diagram add to 180°, the angles are supplementary.

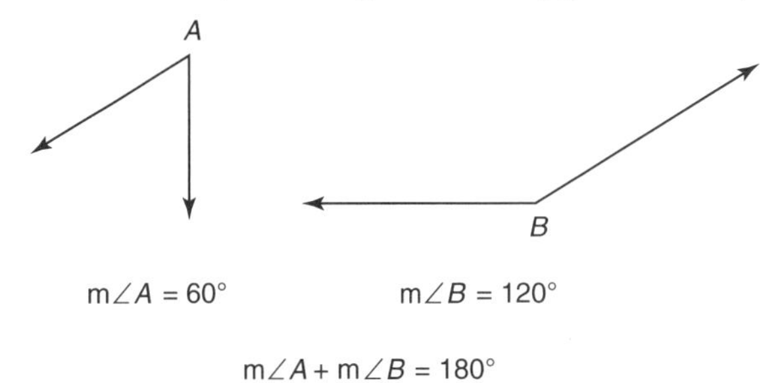


m∠A + m∠B = 180°

## ANGLE BISECTOR

An *angle bisector* cuts an angle into two congruent parts. In the diagram, $\overrightarrow{PQ}$ is the angle bisector of ∠*APB*. The tick marks indicate that the angles are equal.

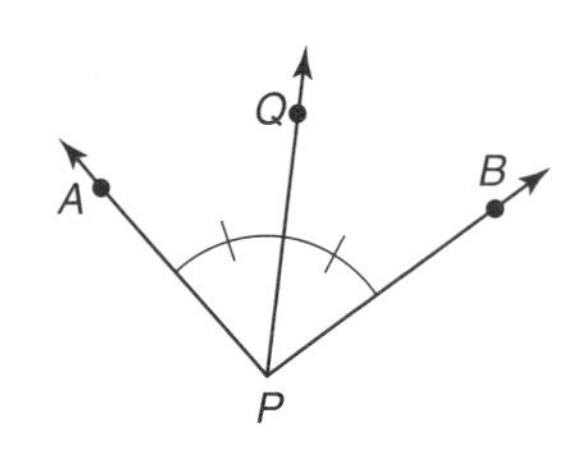


∠*APQ* ≅ ∠*BPQ*

# RELATIONSHIPS BETWEEN INTERSECTING LINES AND ANGLES

When two lines intersect, the angles formed have special relationships.

## Vertical Angles

When two lines intersect, they form four angles. Angles that are opposite each other (not adjacent) are called *vertical angles*. In the diagram, $\angle 1$ and $\angle 3$ are vertical angles. Also, $\angle 2$ and $\angle 4$ are vertical angles. Vertical angles have the same measure. Notice the matching tick marks, which indicate equal angles.

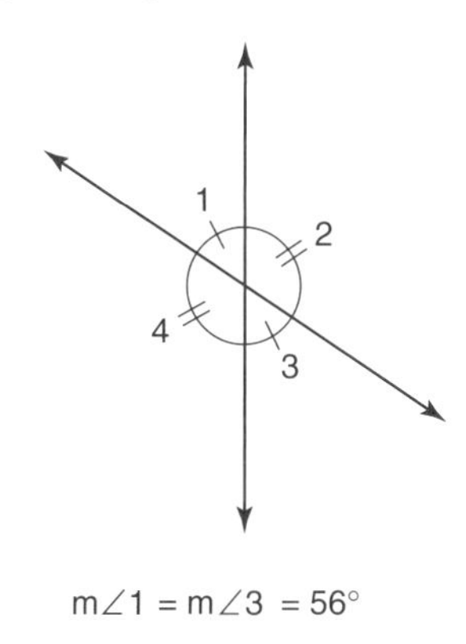


$m\angle 1 = m\angle 3 = 56°$

$m\angle 2 = m\angle 4 = 124°$

## Linear Pair of Angles

A *linear pair of angles* have the same vertex and a common side. The other two sides are opposite rays. In the diagram, $\angle 3$ and $\angle 4$ are a linear pair of angles. Together, their measures add to 180°.

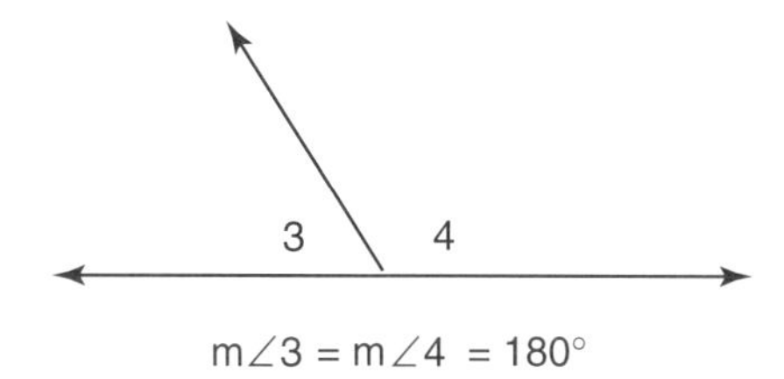


$m\angle 3 = m\angle 4 = 180°$

## RELATIONSHIPS BETWEEN PARALLEL LINES AND ANGLES

When parallel lines are intersected by another line, the intersecting line is called a *transversal*. The pairs of angles formed at these intersections are given special names. In the case of two parallel lines ($\ell_1 \parallel \ell_2$) cut by a transversal ($m$), eight angles are formed, as shown in the diagram.

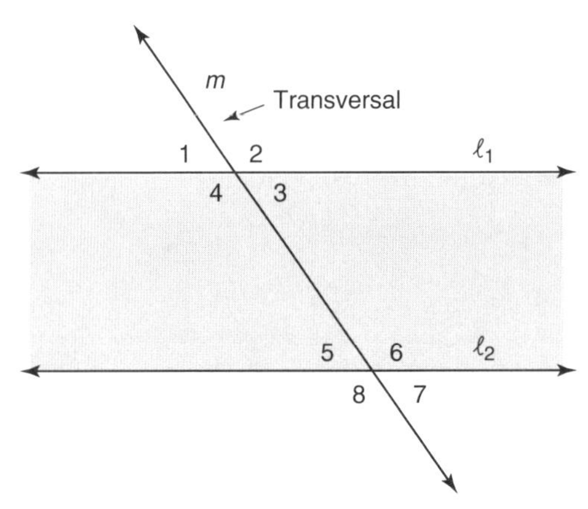


### Interior and Exterior Angles

*Interior angles* are located inside the two parallel lines and, in the diagram shown above, are shaded light gray. These are angles 3, 4, 5, and 6 ($\angle 3$, $\angle 4$, $\angle 5$, $\angle 6$). Angles 1, 2, 7, and 8 are *exterior angles*.

Of the interior angles, $\angle 4$ is an *alternate interior* angle to $\angle 6$, and $\angle 3$ is an *alternate interior* angle to $\angle 5$. Alternate interior angles are congruent. Of the exterior angles, $\angle 1$ is an *alternate exterior* angle to $\angle 7$, and $\angle 2$ is an *alternate exterior* angle to $\angle 8$. Alternate exterior angles are congruent. We say that $\angle 1 \cong \angle 3$ and $\angle 6 \cong \angle 8$.

### Corresponding Angles

In the diagram we have been discussing, angles 1, 2, 3, and 4 *correspond* to angles 5, 6, 7, and 8. Angle 1 corresponds to $\angle 5$, $\angle 2$ corresponds to $\angle 6$, and so on. Corresponding angles are congruent.

## MEASURING AN ANGLE WITH A PROTRACTOR

A *protractor* is an instrument used to measure the number of degrees in an angle. An angle should be adjusted so that its vertex is located as shown in the drawing. The horizontal ray points to a zero, and the other ray points to the degree measurement on one of the scales.

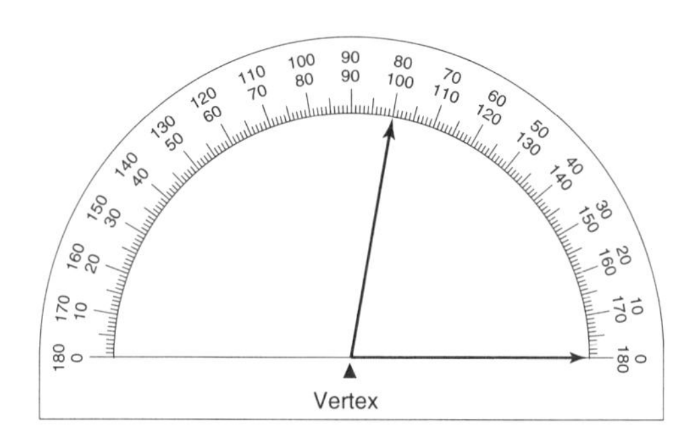


Notice that a protractor has two scales, an upper and a lower scale. To find the measure of an angle, remember that one ray of the angle points toward a zero on one of the scales. This is the scale you should use. The angle shown measures 80°.

On the protractor below, the horizontal ray of the angle points to the zero on the upper scale. Read the degree measurement from the upper scale of the protractor; the measure of the angle is 120°. HINT: This is an obtuse angle, so it *must* measure more than 90°.

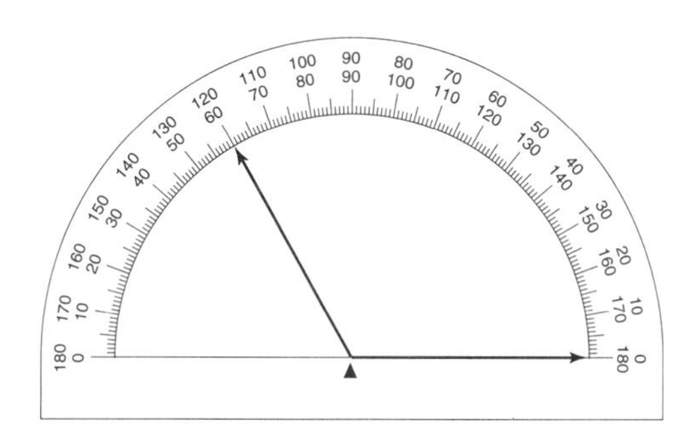

**Example:** In the drawing of a protractor and an angle, the measurement of the angle shown is 45°. This can be found by reading from the lower scale because the horizontal ray of the angle points to the zero on the lower scale. HINT: This is an acute angle, so it *must* measure less than 90°.

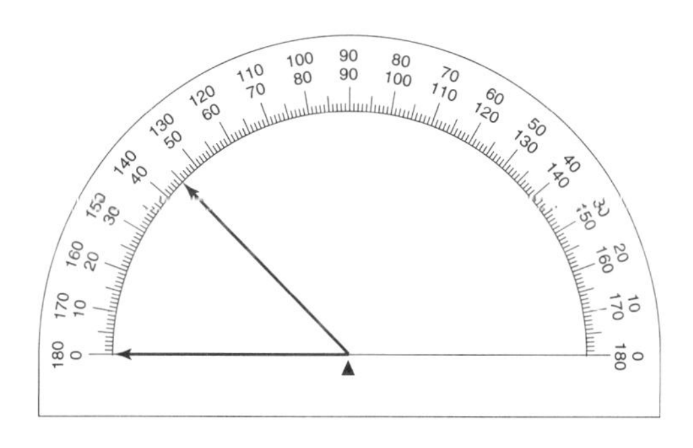

# SAMPLE QUESTIONS

## Angles

1. The interior angles of the octagon shown measure 135°. What is the measure of $\angle VQW$?

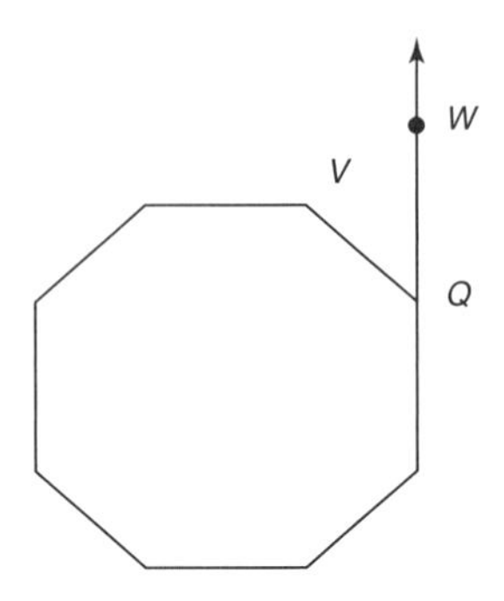


2. The two angles shown below are complementary. Find m$\angle 1$.

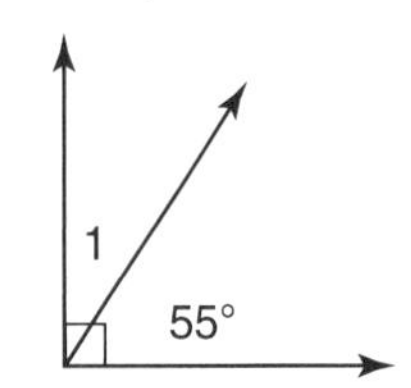


3. In the diagram, $\overrightarrow{BD}$ bisects $\angle ABC$. If m$\angle ABD$ = 40°, find m$\angle DBC$.

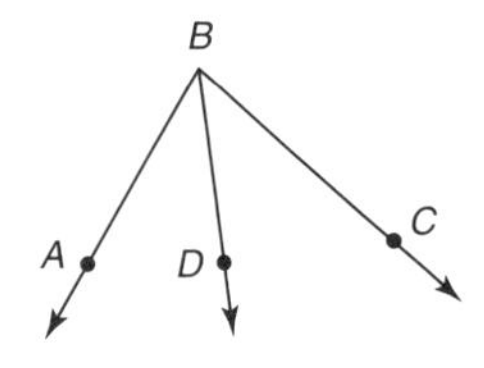


4. In the diagram, which pair of angles are acute?

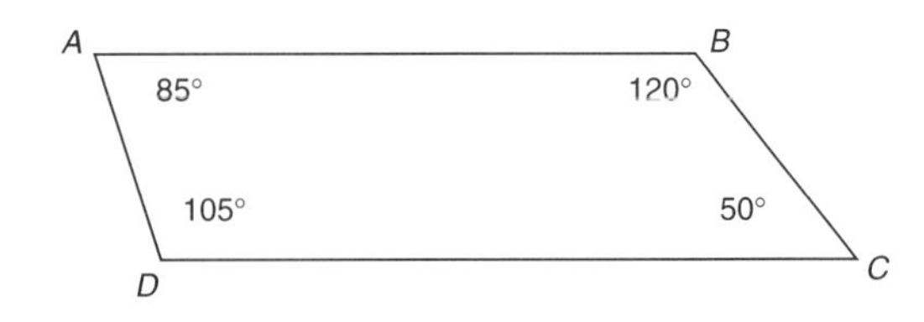


A. $\angle A$, $\angle B$ C. $\angle A$, $\angle C$
B. $\angle B$, $\angle D$ D. $\angle D$, $\angle C$

5. Which type of angle is supplementary to an obtuse angle?

F. acute H. right
G. obtuse I. straight

6. What is the measure of $\angle 2$ in the diagram if $\angle 4$ = 65°?

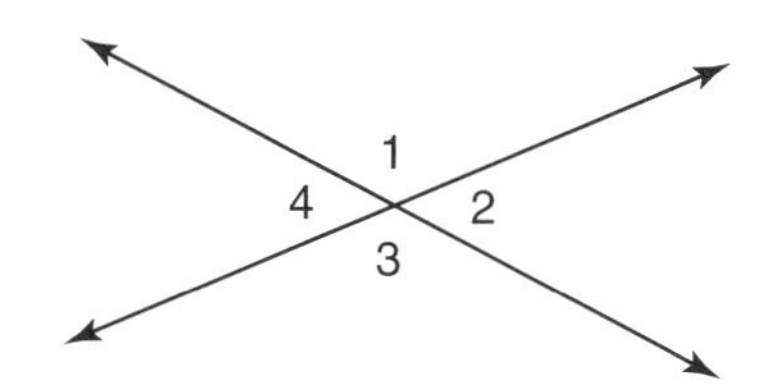

7. Which angle in the diagram is supplementary to $\angle ACB$?

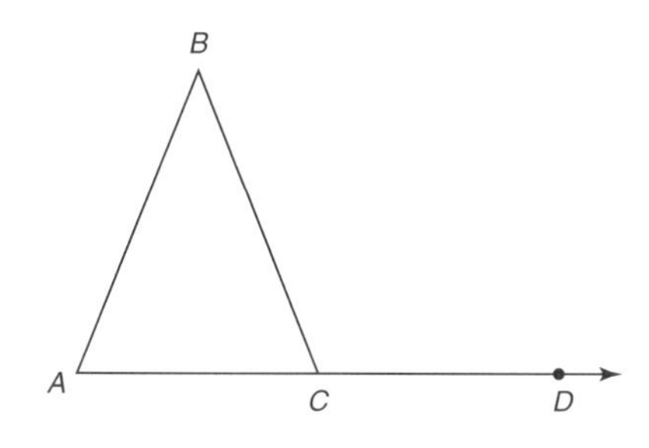


A. $\angle CDA$
B. $\angle BCD$
C. $\angle CAD$
D. $\angle BAC$

8. Find the measure of the angle shown in the diagram.

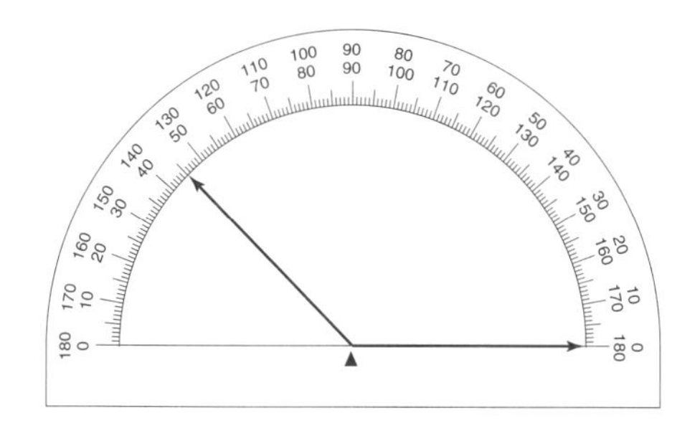

9. In the diagram, what is the relationship between $\angle 1$ and $\angle 2$?

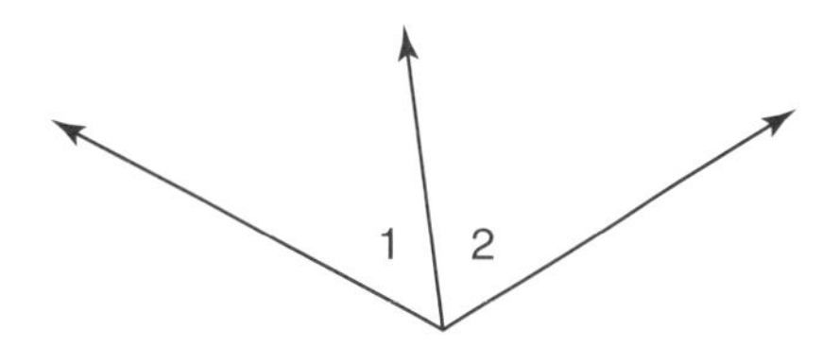


F. congruent
G. linear
H. complementary
I. adjacent

10. Kyle is working with his father to build a bookcase. He is using a miter saw to cut a 45° angle. Which drawing best shows the correct angle cut?

A. 

B. 

C.

D.

## ANSWERS TO SAMPLE QUESTIONS

1. **45.** Angle $VQW$ is part of a linear pair with the other angle measuring 135°. The two angles are supplementary: 180° – 135° = 45°.

2. **35.** Because $\angle 1$ is complementary to a 55° angle, the two angles add to 90°. Subtract: 90° – 55° = 35°.

3. **40.** An angle bisector cuts an angle into two equal parts. This means that $m\angle ABD = m\angle DBC = 40°$.

4. **C.** Both $\angle A$ and $\angle C$ are acute angles. Acute angles measure less than 90°.

5. **F.** Supplementary angles add to 180°. Only an acute angle is small enough to add to an obtuse angle and not exceed 180°.

6. **65.** Angle 2 is vertical to $\angle 4$, so the two angles are equal.

7. **B.** $\angle ACB$ and $\angle BCD$ are a linear pair. Therefore, they are also supplementary.

8. **135.** First, note that this angle is obtuse, so it will measure more than 90°. Then, find the measure of the angle using the upper scale.

9. **I.** Angles 1 and 2 share an endpoint and a side, so they are adjacent. Not enough information is given to allow you to decide whether they are congruent.

10. **C.** A 45° angle is exactly half of a 90° angle. A and D are too small, and B is a 90° angle.

# Lesson 8

## Polygons

A *polygon* is a closed-plane figure made up of straight line segments. A polygon is classified by the number of sides in the figure.

A polygon in which all the sides are the same length is called a *regular polygon.* A common example of a regular polygon is a *square.*

The polygons shown in the table are the ones that appear most commonly on the FCAT.

| Number of Sides | Name of Polygon | Example |
|---|---|---|
| 3 | Triangle | |
| 4 | Quadrilateral | |
| 5 | Pentagon | |
| 6 | Hexagon | |
| 8 | Octagon | |

A polygon is either *concave* or *convex*. The difference between the two types is easy. On a concave polygon, a diagonal can be drawn that goes outside the polygon. A convex polygon has all diagonals on the inside. In the diagrams, the diagonals are shown with dotted lines.

Concave

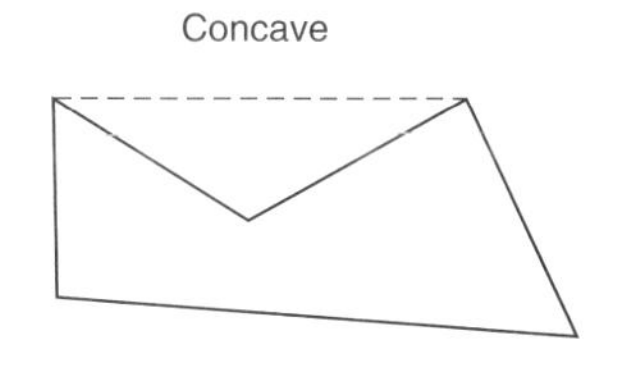

Convex

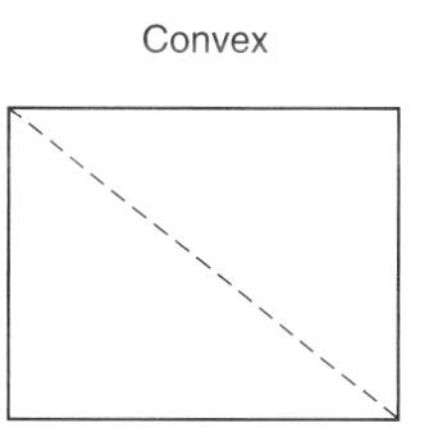

## TRIANGLES

*Triangles* (three-sided polygons) are classified in two ways: by their angle measures and by the number of sides of the same length.

### Triangle Classification by Angles

| Name of Triangle | Type of Angle | Example |
|---|---|---|
| Right | One 90° angle | 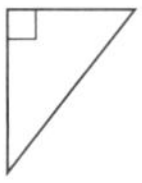 |
| Obtuse | One angle greater than 90° |  |
| Acute | All angles less than 90° | 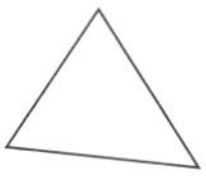 |
| Equiangular | Every angle equal to 60° | 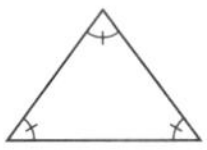 |

### Triangle Classification by Sides

| Name of Triangle | Number of Congruent Sides | Example |
|---|---|---|
| Equilateral | All sides congruent | 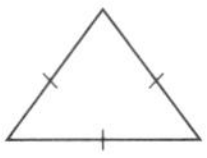 |
| Isosceles | Two sides equal | 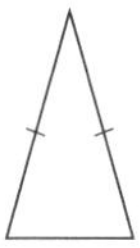 |
| Scalene | No two sides the same | 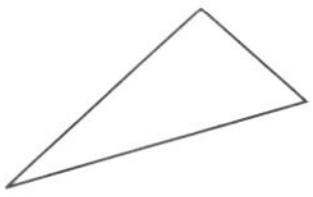 |

A *right scalene triangle* has one right angle and no sides equal.

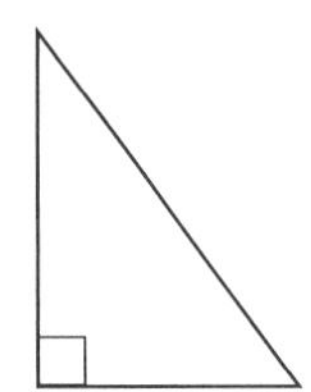

An *obtuse isosceles triangle* has two equal sides, and the angle between them ($\angle B$ in the diagram) is greater than (>) 90°. The other two angles, $\angle A$ and $\angle C$, are equal because the sides opposite them are equal.

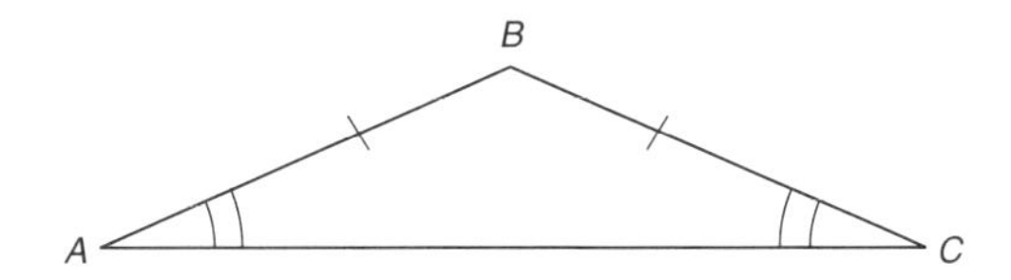


The sum of the angles in a triangle is 180°. Therefore, no triangle can have more than one right angle or one obtuse angle. Also, an equilateral triangle has three 60° angles (180° ÷ 3 = 60°).

## QUADRILATERALS

All polygons with four sides are called *quadrilaterals*. Each quadrilateral has its own special properties, and these properties are used to classify and name the quadrilateral. The sum of the angles in a quadrilateral is 360°.

## PARALLELOGRAMS

This is the group of quadrilaterals that have two sets of parallel sides:

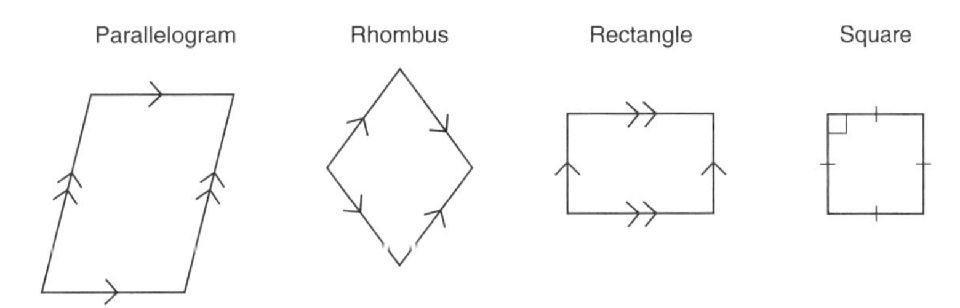


The *square* and *rhombus* have equal sides, while the *parallelogram* and *rectangle* have two sets of parallel sides with different lengths.

Squares and rectangles have right angles.

Parallelograms and rhombi have two sets of supplementary angles.

### Diagonals in Parallelograms

Diagonals in a parallelogram bisect each other but will NOT bisect the angles.

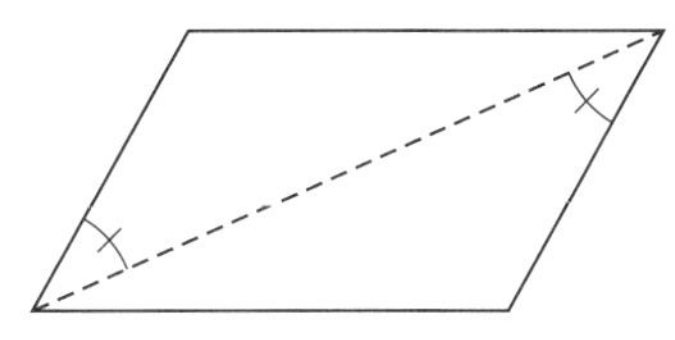

In a rhombus or square, the diagonals are perpendicular:

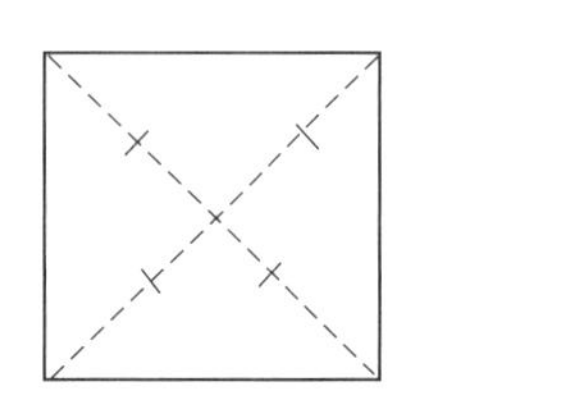

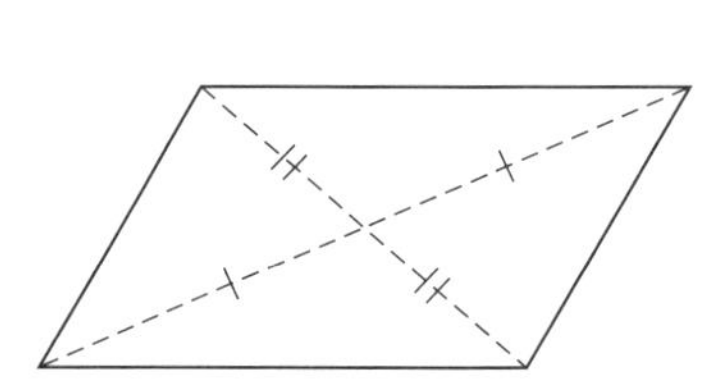

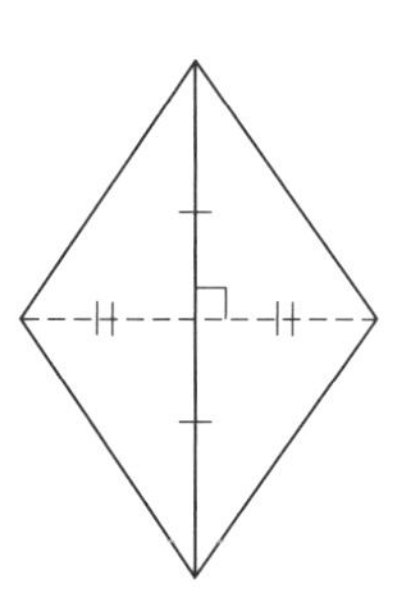

## TRAPEZOID

A *trapezoid* is a quadrilateral with only one set of parallel sides.

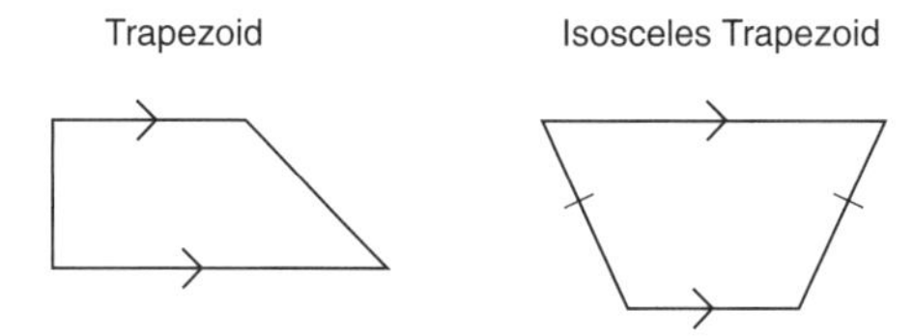


To clarify the relationships among quadrilaterals, a Venn diagram is helpful. In the diagram you can see that a rhombus is actually a very specialized type of parallelogram and that a square is both a specialized rectangle AND a specialized rhombus.

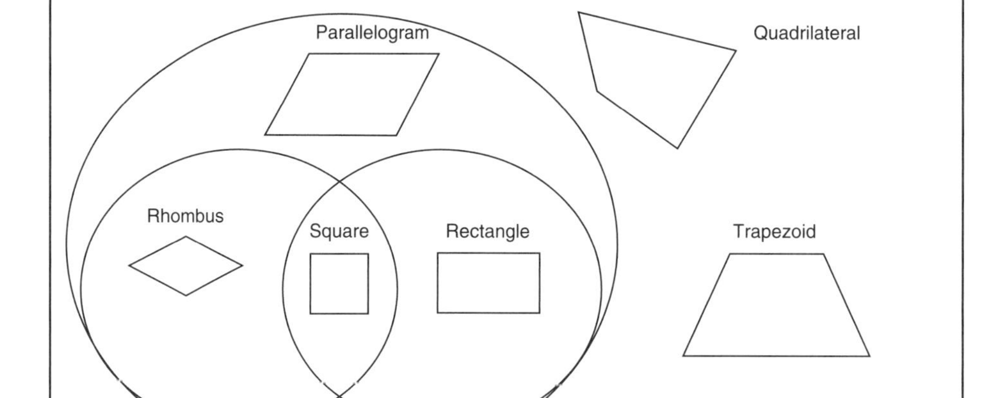

## ANGLES OF POLYGONS

The FCAT Reference Sheet provides formulas to calculate the sum of the angles in a polygon and to find the measure of one angle in a polygon.

**Example 1:** Find the sum of the interior angles of a regular hexagon.

Use the formula from the FCAT Reference Sheet:
$180(n - 2)$, where $n$ = number of sides of the polygon. Hexagons have six sides.

$$180(6 - 2) = 180 \cdot 4 = 720^\circ$$

**Example 2:** Find the measure of one interior angle of a regular pentagon.

Use the formula from the FCAT Reference Sheet:
$\frac{180(n-2)}{n}$, where $n$ = number of sides of the polygon. Pentagons have five sides.

$$\frac{180(5-2)}{5} = \frac{180(3)}{5} = \frac{540}{5} = 108^\circ$$

**Example 3:** Find the measure of one exterior angle of a regular octagon.

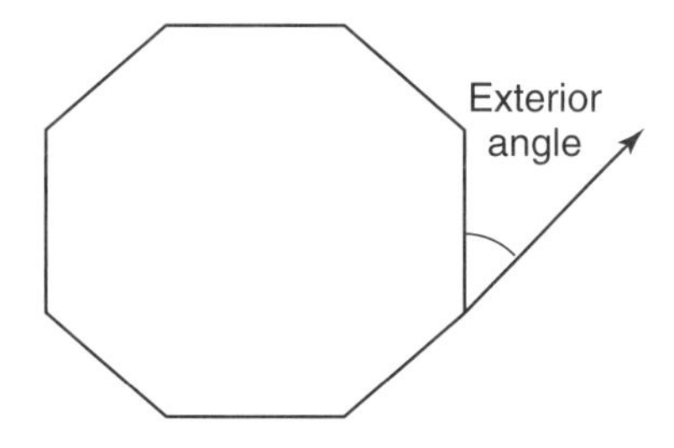


Use the formula from the FCAT Reference Sheet:
$\frac{180(n-2)}{n}$, where $n$ = number of sides of the polygon.

$$\frac{180(8-2)}{8} = \frac{180(6)}{8} = \frac{1080}{8} = 135^\circ$$

Exterior angles and interior angles are supplementary. Therefore, the interior angle (135°) plus the exterior angle will equal 180°. Subtract: 180 – 135 = 45°. The exterior angle is 45°.

## TESSELATIONS

A *tesselation* is a filling up of space with figures permitting no overlaps and no holes. The honeycomb figure shown is tesselated with hexagons. Adjacent angles inside the figure will always add to 360°.

To test this, find the measure of one interior angle of a hexagon:

$$\frac{180(n-2)}{n} = \frac{180(6-2)}{6} = \frac{180(4)}{6} = \frac{720}{6} = 120°.$$

There are three adjacent angles in the honeycomb: $120 \cdot 3 = 360°$. All regular hexagons will tesselate.

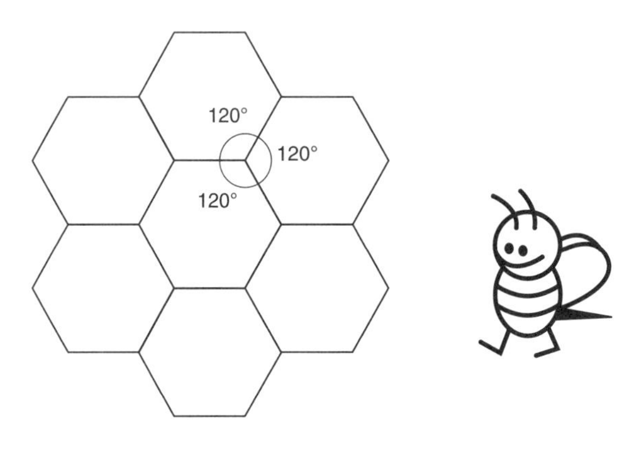


**Example 2:** Will octagons tesselate?

Use the formula: $\frac{180(n-2)}{n} = \frac{180(8-2)}{8} = \frac{180(6-2)}{8} = \frac{180(6)}{8} = \frac{1080}{8} = 135°$

The sum of the adjacent angles of two octagons would be $135 \cdot 2 = 270°$. Three octagons would have an adjacent angle sum of $135 \cdot 3 = 405°$. For a figure to tesselate, the sum of adjacent angles must be exactly 360°. The adjacent angles of two octagons add to less than 360°, and of three octagons to more than 360°; therefore octagons will not tesselate.

# SAMPLE QUESTIONS

## Polygons

1. Use what you know about parallel lines and quadrilaterals to fill in the table by placing a checkmark for each figure that has the property indicated in the first column.

| Property | Parallelogram | Rectangle | Rhombus | Square |
|---|---|---|---|---|
| Opposite sides are parallel (‖). | | | | |
| Opposite sides are congruent (≅). | | | | |
| Opposite angles are ≅. | | | | |
| Diagonals bisect each other. | | | | |
| A diagonal forms two ≅ triangles. | | | | |
| Diagonals are ≅. | | | | |
| A diagonal bisects two angles. | | | | |
| All angles are right angles. | | | | |
| All sides are ≅. | | | | |

2. The pickets in the fence represent what shapes?

A. regular concave pentagons
B. regular convex pentagons
C. irregular concave pentagons
D. irregular convex pentagons

3. If the measure of $\angle 1$ on the envelope shown is 35°, what is the measure of $\angle 2$?

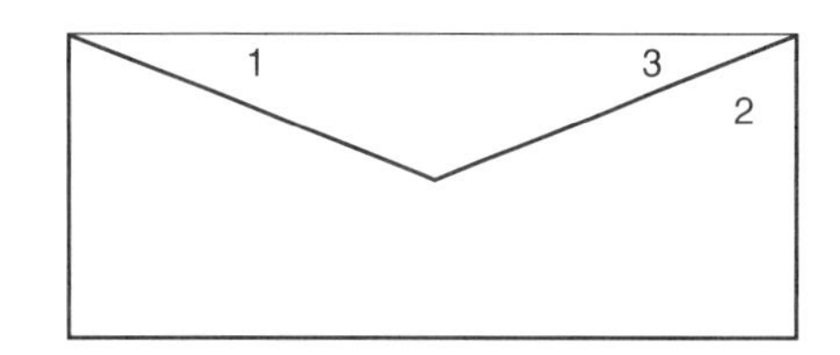


4. Find the measure of $\angle a$ in the parallelogram if $\angle 1 = 115°$.

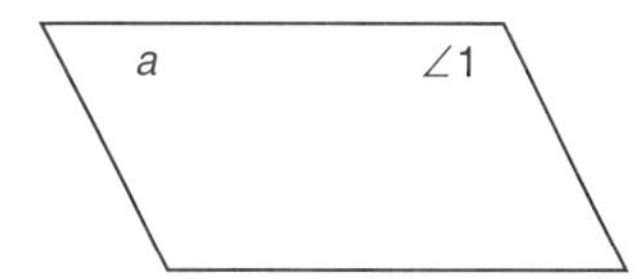


5. For the rhombus shown, which statement is NOT true?

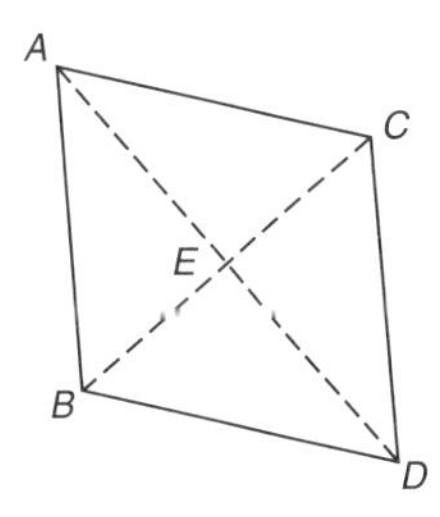


F. $\overline{AB} \cong \overline{AC}$
G. $\angle BED = 90°$
H. $\angle EAC + \angle ECA = 90°$
I. $\angle BAE \cong \angle CAE$

6. What type of angle can be the supplement of an obtuse angle?

   A. right
   B. acute
   C. obtuse
   D. straight

7. Triangle $ABC$ is isosceles. One angle measures 112°. What is the measure of one of the remaining angles?

8. Find the measure of $\angle a$ in the diagram.

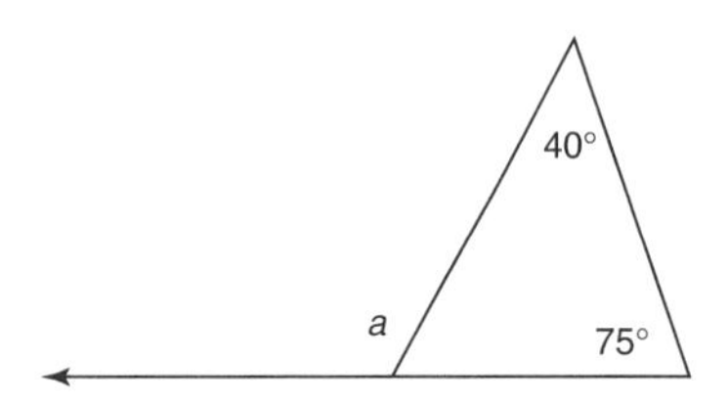


   F. 115°  H. 95°
   G. 105°  I. 65°

9. Takara is attaching wire supports to her tree in preparation for a storm. As shown in the drawing, she attaches the wire 8 feet high on the trunk so that the wire makes a 45° angle with the ground. How many feet from the base of the tree did she anchor the wire?

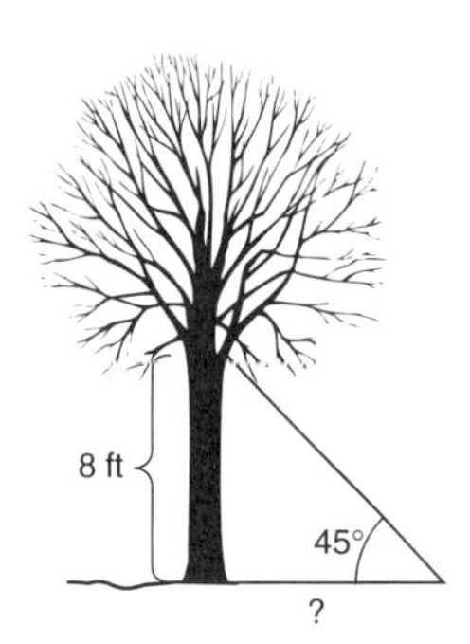


   A. 8  C. 25
   B. 15  D. 45

10. Which of the following figures will NOT tesselate?

    F. triangle  H. rectangle
    G. pentagon  I. hexagon

11. The *median* of a trapezoid is a line drawn parallel to the bases of the trapezoid. The length of the median is the average of the lengths of the two bases. Find the length, in centimeters, of the median of the trapezoid shown.

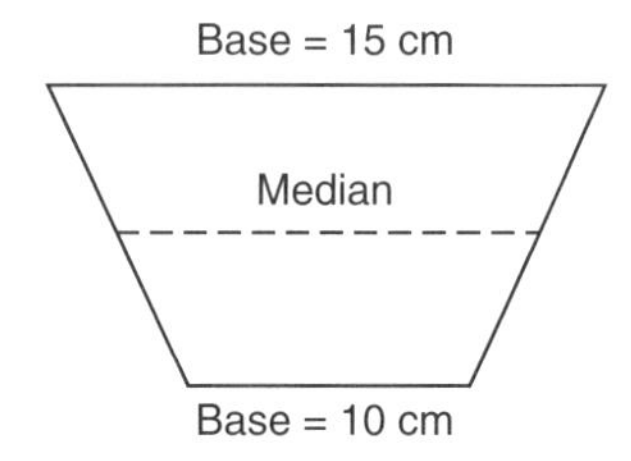


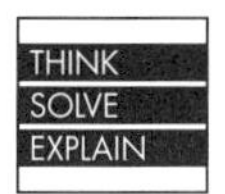


12. Use the figure below to answer the questions.

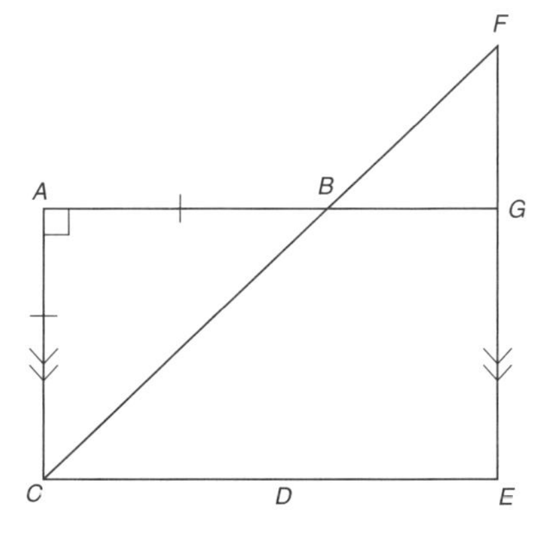


Part A. Find the measure of $\angle ACB$ and $\angle ABC$.

______________________________

______________________________

______________________________

Part B. Find the measure of $\angle CBG$.

______________________________

______________________________

______________________________

THINK
SOLVE
EXPLAIN

**13.** Is it possible for a triangle to have more than one right angle? Explain

______________________________

______________________________

______________________________

**14.** How many equilateral triangles will form a tesselation?

**A.** 12
**B.** 8
**C.** 6
**D.** 3

**15.** A quadrilateral has angles measuring 45°, 102°, and 78°. What is the measure of the fourth angle?

**16.** Which statement is NOT true?

**F.** All rhombi are parallelograms.
**G.** All squares are rectangles.
**H.** Some quadrilaterals are trapezoids.
**I.** Some squares are hexagons.

## ANSWERS TO SAMPLE QUESTIONS

1.

| Property | Parallelogram | Rectangle | Rhombus | Square |
|---|---|---|---|---|
| Opposite sides are parallel (‖). | ✔ | ✔ | ✔ | ✔ |
| Opposite sides are congruent (≅). | ✔ | ✔ | ✔ | ✔ |
| Opposite angles are ≅. | ✔ | ✔ | ✔ | ✔ |
| Diagonals bisect each other. | ✔ | ✔ | ✔ | ✔ |
| A diagonal forms two ≅ triangles. | ✔ | ✔ | ✔ | ✔ |
| Diagonals are ≅. | | ✔ | | ✔ |
| A diagonal bisects two angles. | | | ✔ | ✔ |
| All angles are right angles. | | ✔ | | ✔ |
| All sides are ≅. | | | ✔ | ✔ |

2. **D.**

3. **55.** $\angle 1$ and $\angle 2$ are complementary. Angle 3 is congruent to $\angle 1$.

4. **65.** Adjacent angles in parallelograms are supplementary; 180 – 115 = 65°.

5. **H.** In a rhombus, all sides are equal; therefore F is true. The diagonals of a rhombus are perpendicular to each other; therefore G is true. A diagonal in a rhombus bisects the angle; therefore I is true.

6. **B.** Supplementary angles must add to 180°. Only an acute angle is small enough to add to an obtuse angle for a total of 180°.

7. **34.** If $\triangle ABC$ is isosceles and one angle measures 112°, the sum of the remaining two angles = 180 – 112 = 68°. Since both angles in an isosceles triangle are the same, divide 68 by 2.

8. **F.** The angles in a triangle must add to 180°. The two given angles add to 115°. Then, 180 – 115 = 65°. Angle $a$ + 65 = 180°, and 180 – 65 = 115°. It is no concidence that the exterior angle ($\angle a$) equals the sum of the two interior angles. This is true of any triangle and is the quickest way to find the answer.

9. **A.** Assume that the tree is perpendicular to the ground. Then the other angle must also be at a 45° angle. Two angles with the same measure indicate an isosceles triangle, so the other side of the triangle must be 8 ft.

10. **G.** One interior angle of a pentagon = 108°. Two interior angles = 216°, three interior angles = 324°, and four interior angles = 432°. Since no combination of pentagons will equal 360°, pentagons will not tesselate.

11. **12.5.** Add the base lengths and divide by 2:

$$\frac{10+15}{2} = \frac{25}{2} = 12.5.$$

12. Part A. **45°.** The tick marks on the figure indicate that $\angle ACB \cong \angle ABC$. Since $\triangle ABC$ is an isosceles right triangle, 180 – 90 = 90°. Divide by the two remaining angles; both $\angle ACB$ and $\angle ABC$ = 45°.

    Part B. **135°.** Together, $\angle ABC$ and $\angle CBG$ form a straight angle, so they add to 180°. Then, 180 – 45 = 135°. Therefore, $\angle CBG$ = 135°.

13. It is not possible for a triangle to have more than one right angle because, if a triangle had two right angles, they would add to 180°. Since the sum of the angles in a triangle must be exactly 180°, there would be no way to have a third angle, and the figure formed could not be a triangle.

14. **C.** Equilateral triangles have interior angles of 60°. Six 60° angles are required for a total of 360°.

15. **135.** The sum of the angles in a quadrilateral is found by using the formula 180(4 – 2) = 360°. Add the given angles together, and subtract from 360°:
    360 – (45 + 78 + 102) = 135°.

16. **I.** A square has four sides. A hexagon has six sides.

# Lesson 9

## Powers and Roots

### POWERS

Write $5^6$ in standard form.

*Powers* provide a shorthand way of expressing repeated multiplication. Rather than say $5 \times 5 \times 5 \times 5 \times 5 \times 5$ (expanded notation), we say $5^6$ or *five to the sixth power*. This means six 5's multiplied together. $5^6 = 15{,}625$.

A power such as $5^6$ has a *base* (5) and an *exponent* (6).

$$\text{Base} \longrightarrow 5^{6} \longleftarrow \text{Exponent}$$

The number 25 *written as a power* is $5^2$.

**Example 1:** Write $10 \times 10 \times 10 \times 10$ as a power.

$10^4$

**Example 2:** Write 81 as a power.

$9^2$

**Example 3:** Write $3^3$ in expanded notation.

$3 \times 3 \times 3$

**Example 4:** Write $7^2$ in standard form.

49

Numbers raised to the second power are said to be *squared*. Numbers raised to the third power are *cubed*.

**Example 1:** $4^2$ can be read as *four squared* or *four to the second power.*

**Example 2:** $5^3$ can be read as *five cubed* or *five to the third power.*

## Positive and Negative Exponents

The exponent of a power can be positive or negative. Notice the pattern in the table shown. Compare $10^3$ and $10^{-3}$: $10^3$ equals 1000 and $10^{-3}$ equals $\frac{1}{1000}$. Similarly, $10^2$ equals 100, while $10^{-2}$ equals $\frac{1}{100}$.

**Important note: ANY number to the zero power equals 1!**

| $10^{-3}$ | $10^{-2}$ | $10^{-1}$ | $10^0$ | $10^1$ | $10^2$ | $10^3$ |
|---|---|---|---|---|---|---|
| $\frac{1}{1000}$ | $\frac{1}{100}$ | $\frac{1}{10}$ | 1 | 10 | 100 | 1000 |

**Example 1:** $5^3 = 125$, $5^{-3} = \frac{1}{125}$

**Example 2:** $3^2 = 9$, $3^{-2} = \frac{1}{9}$

## Scientific Notation

*Scientific notation* was developed as a way to write very large or very small numbers without including all the zeros.

**Example:** 350 000 000 000 000 000 000 000 is written in scientific notation as $3.5 \times 10^{23}$.

All numbers written in scientific notation must have the following form:

$$a \times 10^b$$

The number represented by the letter $a$ must be 1 or more, but less than 10. This number is multiplied by a positive or negative power of 10. The exponent of 10 is represented by $b$. Positive exponents result in numbers larger than 1 while negative exponents indicate numbers smaller than 1.

**Example 1:** $4.5 \times 10^5 = 4.5 \times 10 \times 10 \times 10 \times 10 \times 10 = 450{,}000.$

Multiplying by 10 five times moves the decimal point in 4.5 five places to the right. Four zeros are needed to fill in.

**Example 2:** $3.1 \times 10^{-4} = 3.1 \times \frac{1}{10} \times \frac{1}{10} \times \frac{1}{10} \times \frac{1}{10} = 0.00031.$

Multiplying by $\frac{1}{10}$ four times moves the decimal point in 3.1 four places to the left. Three zeros are needed to fill in.

## SQUARE ROOTS

Between what two consecutive integers is $\sqrt{50}$?

Since $\sqrt{50}$ is approximately 7.1, it is between the integers 7 and 8.

To understand square roots, it helps to take a moment to discuss some of the numbers that are perfect squares: 1, 4, 9, 16, 25, 36, 49, 64, 81, and 100. These numbers are called *perfect squares* because they represent whole numbers that have been *squared*.

| | |
|---|---|
| $1^2 = 1$ | $6^2 = 36$ |
| $2^2 = 4$ | $7^2 = 49$ |
| $3^2 = 9$ | $8^2 = 64$ |
| $4^2 = 16$ | $9^2 = 81$ |
| $5^2 = 25$ | $10^2 = 100$ |

The symbol $\sqrt{\ }$ is called the *square root* symbol or a *radical*. We say that the *square root* of 25 ( $\sqrt{25}$ ) equals 5 because $5 \times 5$ equals 25. You can think of taking a square root as *undoing* a square:

| | |
|---|---|
| $\sqrt{1} = \sqrt{1^2} = \sqrt{1 \times 1} = 1$ | $\sqrt{36} = \sqrt{6^2} = \sqrt{6 \times 6} = 6$ |
| $\sqrt{4} = \sqrt{2^2} = \sqrt{2 \times 2} = 2$ | $\sqrt{49} = \sqrt{7^2} = \sqrt{7 \times 7} = 7$ |
| $\sqrt{9} = \sqrt{3^2} = \sqrt{3 \times 3} = 3$ | $\sqrt{64} = \sqrt{8^2} - \sqrt{8 \times 8} = 8$ |
| $\sqrt{16} = \sqrt{4^2} = \sqrt{4 \times 4} = 4$ | $\sqrt{81} = \sqrt{9^2} = \sqrt{9 \times 9} = 9$ |
| $\sqrt{25} = \sqrt{5^2} = \sqrt{5 \times 5} = 5$ | $\sqrt{100} = \sqrt{10^2} = \sqrt{10 \times 10} = 10$ |

Not all square roots represent whole numbers: $\sqrt{2} = 1.41421\ldots$. In fact, most square roots have to be rounded.

There is a square root key on your FCAT calculator. On it, you will see the $\sqrt{\ }$ symbol. To take the square root of a number, enter the number first, then hit the square root key.

# SAMPLE QUESTIONS

## Powers and Square Roots

1. The speed of light is approximately 186,000 miles per second. Which of the following expresses this number in scientific notation?

   A. $1.86 \times 10^3$
   B. $18.6 \times 10^4$
   C. $1.86 \times 10^5$
   D. $18.6 \times 10^3$

2. The average toilet is flushed approximately $12^5$ times over its lifetime. Which expression means the same as $12^5$?

   F. $12 \times 5$
   G. $12 + 12 + 12 + 12 + 12$
   H. $12 \times 12 \times 12 \times 12 \times 12$
   I. $(12 \times 5) + (12 \times 5) + (12 \times 5) + (12 \times 5) + (12 \times 5)$

3. Place in order from smallest to largest: $-3, 10^{-2}, \sqrt{2}, 5^0$.

   A. $-3, 10^{-2}, 5^0, \sqrt{2}$
   B. $10^{-2}, -3, \sqrt{2}, 5^0$
   C. $-3, \sqrt{2}, 5^0, 10^{-2}$
   D. $-3, 5^0, 10^{-2}, \sqrt{2}$

4. Between what two consecutive integers is $5\sqrt{2}$ located?

   F. 6 and 7
   G. 7 and 8
   H. 7 and 7.1
   I. 8 and 9

5. Fill in the blanks in the table.

| Standard Form | Scientific Notation | Exponent | Square Root |
|---|---|---|---|
| 1 | | | |
| 36 | | | |
| 81 | | | |
| 100 | | | |

6. $\sqrt{75}$ is closest to which perfect square?

   A. 81
   B. 64
   C. 9
   D. 8

7. $10^6$ is how many times larger than $10^3$?

8. Add $8^2 + 5^2$.

   F. 89
   G. 41
   H. 36
   I. 18

9. Which of the following pairs of numbers are equivalent?

   A. 81 and $2^8$
   B. $2^4$ and 32
   C. $3.5 \times 10^4$ and 35,000
   D. $3.5 \times 10^4$ and 350,000

10. The average distance of Mars from the Sun is $2.32 \times 10^8$. Which number represents this distance in standard notation?

    F. 2,320,000
    G. 23,200,000
    H. 232,000,000
    I. 23,200,000,000

11. The "hairy-winged" beetles are the smallest known member of the Trichopterygidae family. They measure $2 \times 10^{-2}$ centimeter in length. What is this number in standard notation?

12. The number 0.000007 written in scientific notation is

    A. $7 \times 10^{-6}$
    B. $7 \times 10^{-4}$
    C. $7 \times 10^{-5}$
    D. $7 \times 10^{-3}$

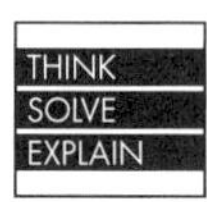


13. Part A. Show that the sum of the squares of 3 and 4 equals the square of 5.

    ______________________________

    ______________________________

    ______________________________

    Part B. In Part A above, 3, 4, and 5 are consecutive integers. Is it always true that the sum of the squares of two consecutive integers equals the square of the next integer? Show or explain by giving an example.

    ______________________________

    ______________________________

    ______________________________

## ANSWERS TO SAMPLE QUESTIONS

1. **C.** Place the decimal between 1 and 8. The decimal point moved five places, so the exponent is 5. In scientific notation, the speed of light is approximately $1.86 \times 10^5$ mi/sec.

2. **H.** The expression $12^5$ means five 12's multiplied together.

3. **A.** $-3 = -3$; $10^{-2} = 0.01$; $5^0 = 1$; $\sqrt{2} = 1.414\ldots$.

4. **G.** Find $\sqrt{2}$, and multiply by 5: $\sqrt{2} = 1.414 \ldots \times 5 = 7.0$ approximately. Answer H is incorrect because 7.1 is not an integer.

5.

| Standard Form | Scientific Notation | Exponent | Square Root |
|---|---|---|---|
| 1 | $1 \times 10^0$ | $1^0$, $1^2$, etc. | $\sqrt{1} = 1$ |
| 36 | $3.6 \times 10^1$ | $6^2$ | $\sqrt{36} = 6$ |
| 81 | $8.1 \times 10^1$ | $9^2$ | $\sqrt{81} = 9$ |
| 100 | $1 \times 10^2$ | $10^2$ | $\sqrt{100} = 10$ |

6. **C.** $\sqrt{75} = 8.66$, which is closest to 9.

7. **1000.** If you multiply $10^3$ by 10 (1000) three times, the result is 1,000,000 or $10^6$.

8. **F.** $8^2 + 5^2 = 64 + 25 = 89$

9. **C.** $3.5 \times 10^4 = 35{,}000$.

10. **H.** Start with 2.32 and move the decimal point eight places to the right (positive power of 10). In standard notation, the distance is 232,000,000 km.

11. **0.02.** Start with 2.0 and move the decimal point two places to the left (negative power of 10).

12. **A.** Start with 0.000007 and move the decimal point six places to the right until it is after the 7. Multiply by $10^{-6}$.

13. Part A. $3^2 + 4^2 = 5^2$ because $3^2 = 9$, $4^2 = 16$, and $5^2 = 25$; $9 + 16 = 25$.

    Part B. **No.** If it were true, then $1^2 + 2^2$ would equal $3^2$. $1^2 + 2^2$ equals $1 + 4 = 5$. The sum of $1^2$ and $2^2$ does not equal $3^2$, which equals 9.

# Lesson 10

# The Pythagorean Theorem and Special Right Triangles

## THE PYTHAGOREAN THEOREM

At exactly 3 o'clock the hands on a clock form a right triangle. If the hour hand (short) is 5 centimeters long, and the minute hand (long) is 12 centimeters long, how far apart are the ends of the hands?

Use the formula $a^2 + b^2 = c^2$.

| | |
|---|---|
| $5^2 + 12^2 = c^2$ | Substitute the values given in the question. |
| $25 + 144 = c^2$ | Simplify the powers. |
| $169 = c^2$ | Add. |
| $13 = c$ | Take the square root of each side. |

The ends of the hands are 13 cm apart.

The Pythagorean theorem is used to find the length of one side of a right triangle when the lengths of the other two sides are known. The theorem says that in any right triangle, the square of the length of the hypotenuse equals the sum of the squares of the lengths of the other two sides.

The theorem is written in algebraic form as $a^2 + b^2 = c^2$.

This formula is given to you on the FCAT Reference Sheet and can be very useful for finding the length of a side of a right triangle.

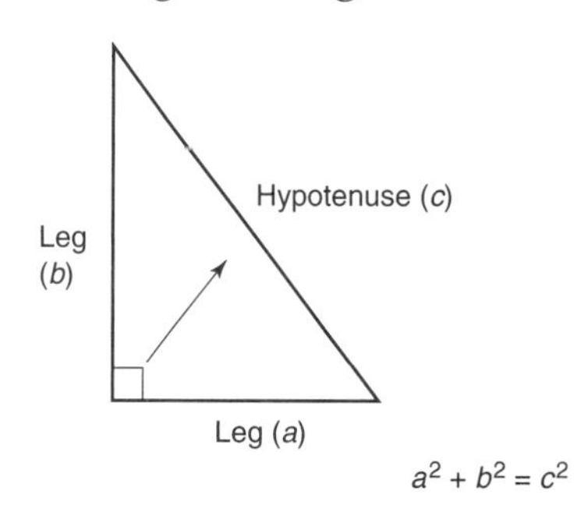

The *hypotenuse* is the longest side of a right triangle. It can be located easily because it is directly opposite the right angle. The remaining two sides of the triangle are called *legs*. The $c$ in the Pythagorean theorem refers to the hypotenuse; $a$ and $b$ represent the legs.

**Example 1:** Find the length of the hypotenuse in the triangle shown.

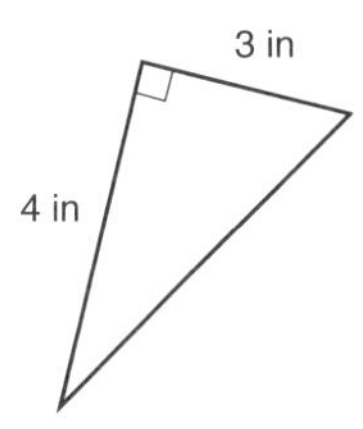


Use the formula $a^2 + b^2 = c^2$.

| | |
|---|---|
| $3^2 + 4^2 = c^2$ | Substitute the values given in the diagram. |
| $9 + 16 = c^2$ | Simplify the powers. |
| $25 = c^2$ | Add. |
| $5 = c$ | Take the square root of each side. |

The length of the hypotenuse is 5 in.

**Example 2:** Find, to the nearest tenth of a centimeter, the missing length in the triangle shown.

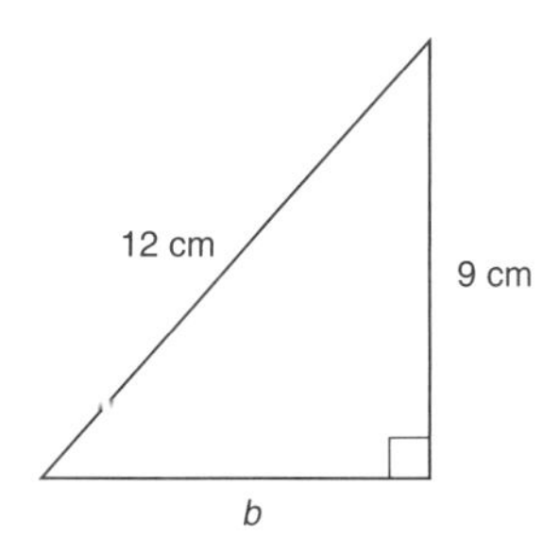


Notice that in this example you are solving for the length of a leg, NOT for the length of the hypotenuse.

Use the formula $a^2 + b^2 = c^2$.

| | |
|---|---|
| $9^2 + b^2 = 12^2$ | Substitute the values given in the diagram. |
| $81 + b^2 = 144$ | Simplify the powers. |
| $b^2 = 144 - 81$ | Subtract. |
| $b^2 = 63$ | Take the square root of each side. |
| $b \approx 7.9$ | The length of the leg is approximately ($\approx$) 7.9 cm. |

**Example 3:** A baseball diamond is actually a square. In the baseball diamond shown, what is the distance, to the nearest foot, from second base to home plate?

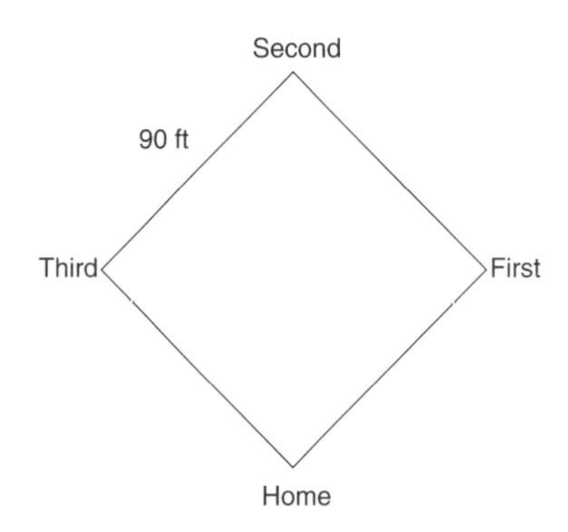


Draw a line from home to second base. This divides the square into two right triangles with equal legs.

Use the formula $a^2 + b^2 = c^2$.

| | |
|---|---|
| $90^2 + 90^2 = c^2$ | Substitute the value given in the diagram. |
| $8100 + 8100 = c^2$ | Simplify the powers. |
| $16200 = c^2$ | Add. |
| $127 = c$ | Take the square root of each side. |

The distance from second base to home plate is approximately 127 ft.

## SPECIAL RIGHT TRIANGLES

There are two special right triangles you should be familiar with: 45-45-90 and 30-60-90. The numbers 45-45-90 and 30-60-90 refer to measurements of the angles in these triangles.

### 45-45-90 Right Triangles

A *45-45-90* triangle has two 45-degree angles and one 90-degree angle, as shown in the diagram. Because this triangle has two congruent angles, the sides opposite these angles are also congruent, making this an *isoceles right triangle*. Both legs of the triangle have the same measure, and the hypotenuse of a 45-45-90 triangle is always equal to $\sqrt{2}$ times the length of a leg.

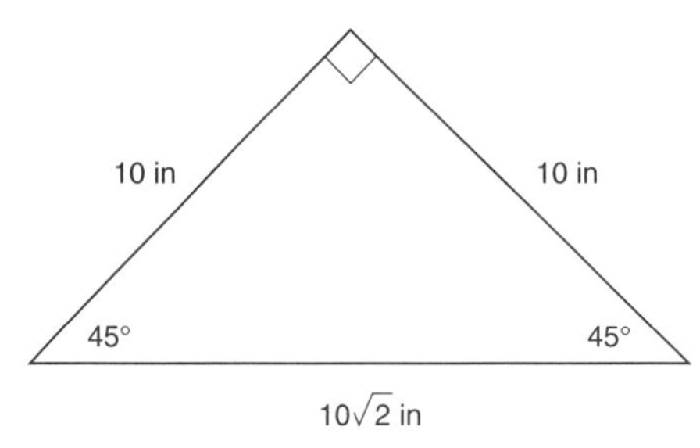

**Example:** The triangle shown is an isosceles right triangle. Find the length of the hypotenuse.

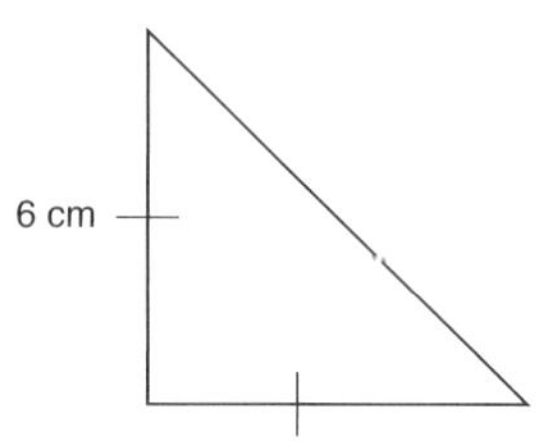


This is a 45-45-90 triangle. The hypotenuse is $6\sqrt{2}$ cm long.

## 30-60-90 Right Triangles

In a *30-60-90* triangle, the length of the hypotenuse is twice the length of the shortest leg (which is opposite the 30-degree angle), as shown. The other leg is always equal to $\sqrt{3}$ times the shortest leg.

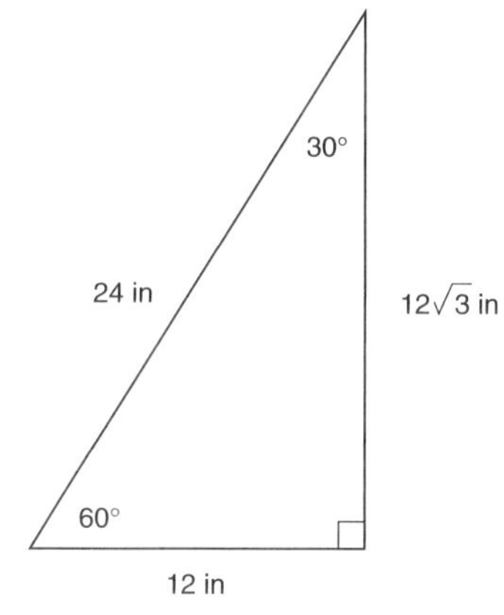


**Example:** Find the length of the shortest leg of the scalene right triangle below.

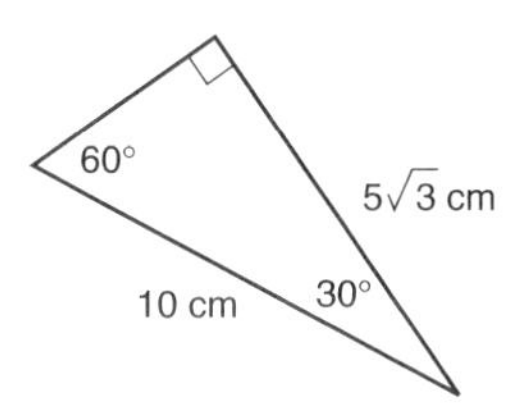


The length of the shortest leg in a 30-60-90 triangle is exactly half the length of the hypotenuse, or 5 cm.

# SAMPLE QUESTIONS

## The Pythagorean Theorem and Special Right Triangles

1. Two neighbors, Mr. Algarin and Mr. Bess, want to find the distance between their two houses, which are situated on opposite sides of a lake, as shown in the drawing. Mr. Bess measured the distance from his house to Jack's shack and found it to be 800 yards. Mr. Algarin found that the distance from his house to Jack's shack was 1200 yards. What is the distance, to the nearest yard, across the lake from Mr. Algarin's house to Mr. Bess's house?

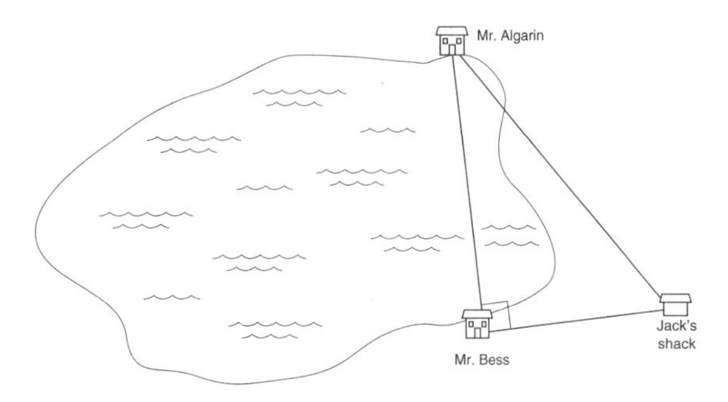


2. Using the Pythagorean theorem, fill in the missing values in the table. Round to the nearest tenth where necessary.

| $a$ | $b$ | $c$ |
|---|---|---|
| 6 | 8 | |
| | 1 | 3 |
| 4 | | 8 |
| 2 | | 4 |
| | 15 | 30 |
| 10 | 12 | |

3. Find the distance, to the nearest inch, from point $A$ to point $C$ in the rectangular prism shown.

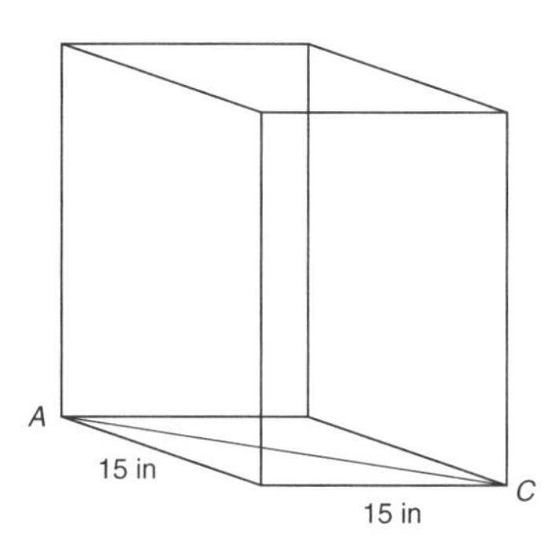


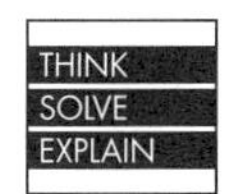


4. The telephone company is attaching a support wire to a telephone pole to provide extra support against an upcoming storm. The wire is attached 20 feet high on the pole so that the wire makes a 45° angle with the ground, as shown in the diagram.

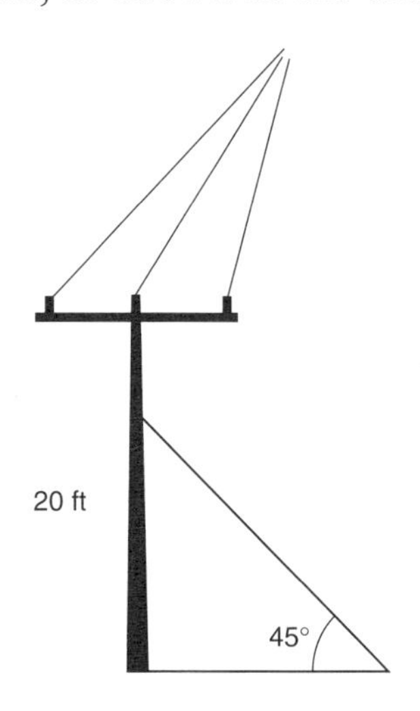

Part A. How many feet from the base of the telephone pole should the company anchor the wire? Explain why.

______________________________

______________________________

______________________________

Part B. How long should the wire be? Explain how you arrived at your answer, or show your work.

______________________________

______________________________

______________________________

5. A 16-foot ladder is placed 4 feet from the base of a building. How far up the building does the ladder reach? Round your answer to the nearest foot.

THINK
SOLVE
EXPLAIN

6. A ship leaves port and sails 12 kilometers west and then 21 kilometers north. How far is the ship from port?

Part A. Draw and label a diagram that represents this situation.

Part B. How far is the ship from port? Show your work, or explain how you arrived at your answer.

______________________________

______________________________

______________________________

7. The window of a burning building is 15 feet above the ground. The base of a 15-foot ladder is placed 10 feet from the building. The ladder does not reach the window. To the nearest foot, how far away is the top of the ladder from the window?

8. A rectangle is 6 centimeters wide and 36 centimeters long. To the nearest tenth of a centimeter, how long is its diagonal?

A. 36.4
B. 36.5
C. 37
D. 38

9. A skateboard ramp rises 4 yards over a horizontal distance of 10 yards, as shown. To the nearest whole yard, how long is the ramp?

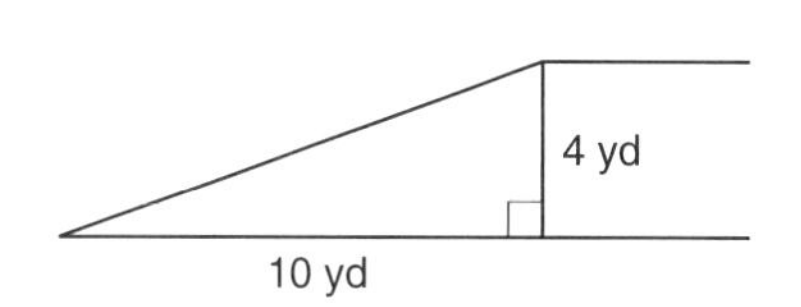

**10.** When Jorge has let out 50 yards of kite string, he notices that his kite is directly above Maria, as shown below. If Jorge is 30 yards from Maria, how high, in yards, is the kite?

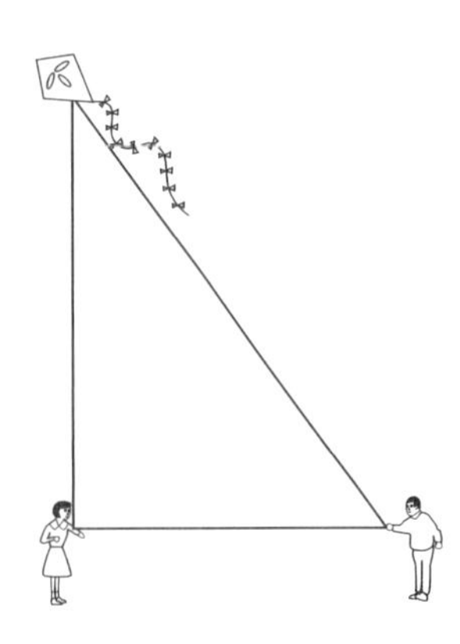

**11.** What is the height, in inches, of the parallelogram shown in the drawing below?

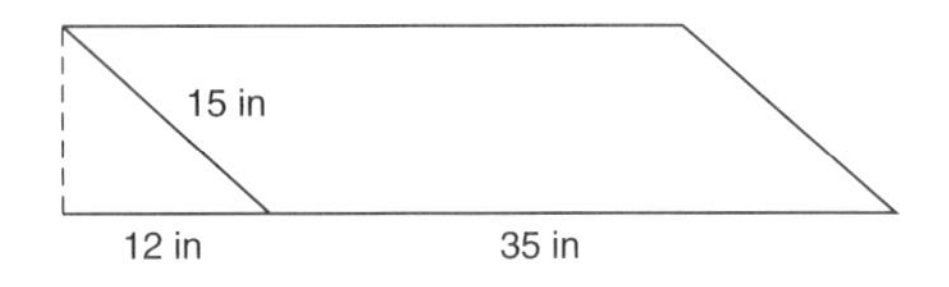


## ANSWERS TO SAMPLE QUESTIONS

1. **894.** Use the formula $a^2 + b^2 = c^2$.

| | |
|---|---|
| $800^2 + b^2 = 1200^2$ | Substitute the values given in the question. |
| $640{,}000 + b^2 = 1{,}440{,}000$ | Simplify the powers. |
| $b^2 = 1{,}440{,}000 - 640{,}000$ | Subtract. |
| $b^2 = 800{,}000$ | |
| $b \approx 894$ | Take the square root of each side. |

The distance is approximately 894 yd.

2.

| *a* | *b* | *c* |
|---|---|---|
| 6 | 8 | 10 |
| 2.8 | 1 | 3 |
| 4 | 6.9 | 8 |
| 2 | 3.5 | 4 |
| 26 | 15 | 30 |
| 10 | 12 | 15.6 |

3. **21.** Use the formula: $a^2 + b^2 = c^2$.

| | |
|---|---|
| $15^2 + 15^2 = c^2$ | Substitute the values given in the diagram. |
| $225 + 225 = c^2$ | Simplify the powers. |
| $450 = c^2$ | Add. |
| $21 \approx c$ | Take the square root of each side. |

4. Part A. **20.** The wire should be attached 20 ft away because this is a 45-45-90 triangle and the two legs are the same length.
Part B. **About 28 ft.** $20^2 + 20^2 = 800$. Take the square root of 800: $\sqrt{800} \approx 28$.

5. **15 ft.** Use the formula: $a^2 + b^2 = c^2$.

| | |
|---|---|
| $4^2 + b^2 = 16^2$ | Substitute the values given in the question. |
| $16 + b^2 = 256$ | Simplify the powers. |
| $16 + b^2 - 16 = 256 - 16$ | Subtract 16 from both sides. |
| $b^2 = 240$ | Simplify. |
| $b = \sqrt{240} = 15.49$ ft | Take the square root of each side. |

6. Part A.

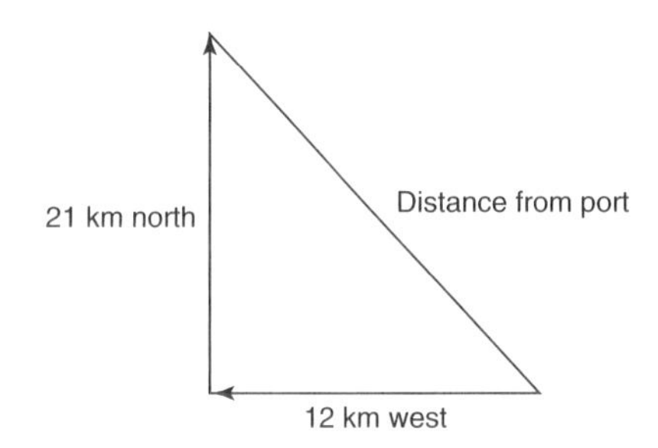


Part B. **24 km.** $12^2 + 21^2 = 144 + 441 = 585$; $\sqrt{585} \approx 24$.

7. **4.** The 15-ft ladder will reach 11.2 ft high, about 4 ft short of 15 ft.

Use the formula $a^2 + b^2 = c^2$.

| | |
|---|---|
| $10^2 + b^2 = 15^2$ | Substitute the values given in the question. |
| $100 + b^2 = 225$ | Simplify the powers. |
| $b^2 = 225 - 100$ | Subtract. |
| $b^2 = 125$ | |
| $b \approx 11.2$ | Take the square root of each side. |

The ladder reaches approximately 11 ft high.

8. **A.** Use the formula $a^2 + b^2 = c^2$.

| | |
|---|---|
| $6^2 + 36^2 = c^2$ | Substitute the values given in the question. |
| $36 + 1296 = c^2$ | Simplify the powers. |
| $1332 = c^2$ | Add. |
| $36.4 \approx c$ | Take the square root of each side. |

9. **11.** Use the formula $a^2 + b^2 = c^2$.

| | |
|---|---|
| $10^2 + 4^2 = c^2$ | Substitute the values given in the diagram. |
| $100 + 16 = c^2$ | Simplify the powers. |
| $116 = c^2$ | Add. |
| $11 \approx c$ | Take the square root of each side. |

10. **40.** Use the formula $a^2 + b^2 = c^2$.

| | |
|---|---|
| $a^2 + 30^2 = 50^2$ | Substitute the values given in the question. |
| $a^2 + 900 = 2500$ | Simplify the powers. |
| $a^2 = 2500 - 900$ | Subtract. |
| $a = 40$ | Take the square root of each side. |

11. **9.** The height of a polygon is perpendicular to the base. The dotted line in the diagram represents the height of the parallelogram shown.

Use the formula $a^2 + b^2 = c^2$.

| | |
|---|---|
| $a^2 + 12^2 = 15^2$ | Substitute the values given in the diagram. |
| $a^2 + 144 = 225$ | Simplify the powers. |
| $a^2 = 225 - 144$ | Subtract. |
| $a^2 = 81$ | |
| $a = 9$ | Take the square root of each side. |

# Lesson 11

# Perimeters and Areas of Polygons

## PERIMETER

Find the perimeter of the square shown.

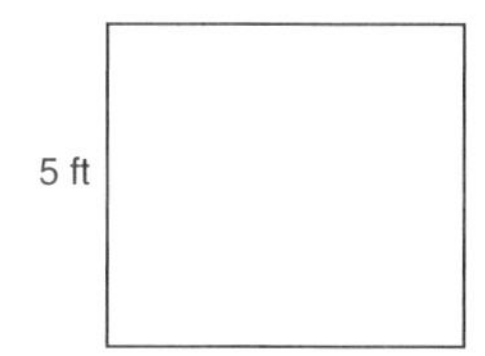


The *perimeter* of a polygon is a measurement around the outside of the figure. In a square, all four sides have the same length. You can add 5 four times, or you can multiply 5 by 4. Either way, the perimeter is 20 feet.

**Example 1:** A farmer's rectangular field measures 1350 feet long by 820 feet wide, as shown. How much fencing will the farmer need to completely enclose the field?

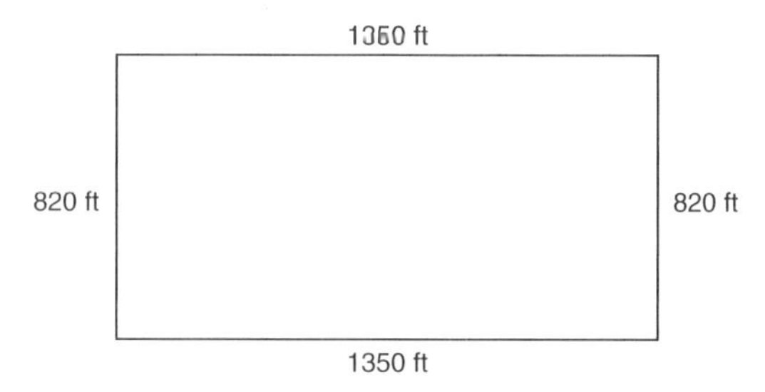


You should draw and label a diagram of any problem that involves perimeter, area, or volume. In the case of this example, the diagram will remind you to add all four sides: 1350 + 1350 + 820 + 820. An alternative is to use the formula

$$\text{Perimeter} = 2(1350) + 2(820).$$

Either way, the perimeter is 4340 ft.

**Example 2:** Find the perimeter of an isosceles triangle with base measuring 11.6 meters and legs measuring 6.5 meters.

Because this triangle is isosceles, two legs have equal measure. Draw an isosceles triangle and label the sides, as shown. Add: 11.6 + 6.5 + 6.5 = 24.6. The triangle's perimeter is 24.6 m.

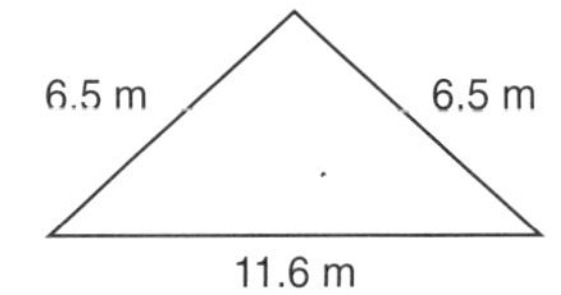


**Example 3:** Find the perimeter of an equilateral pentagon with side measuring 10.02 centimeters.

All five sides are equal in an equilateral pentagon. Draw a pentagon and label it, as shown. Then find the perimeter by multiplying 10.02 by 5. The perimeter equals 50.1 cm.

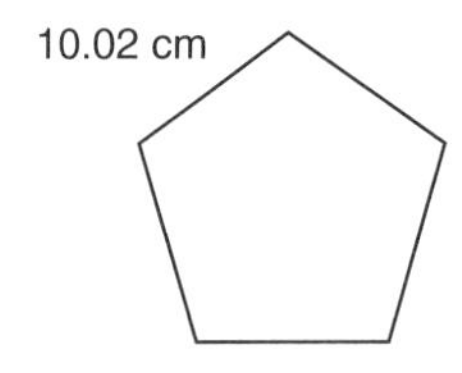


**Example 4:** Find the perimeter of the figure shown.

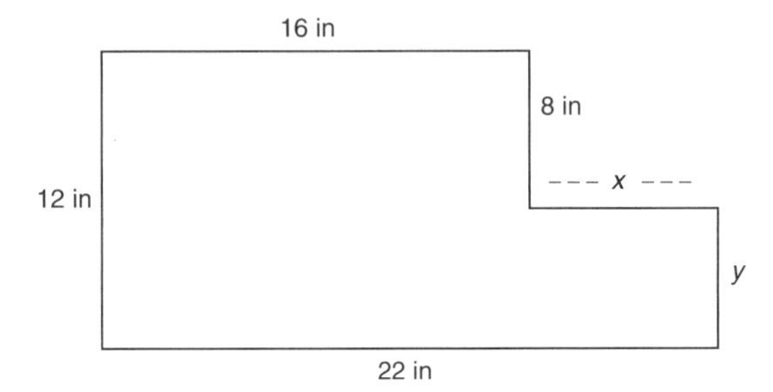


This six-sided figure is an irregular hexagon. To find the perimeter, you must first find the lengths of all six sides.

Find the length of side $x$ by subtracting: 22 – 16 = 6 in. Then find the length of side $y$ by subtracting: 12 – 8 = 4 in. Finally, add all sides to find the perimeter: 16 + 8 + 6 + 4 + 22 + 12 = 68 in.

# AREA

*Area* represents the amount of space inside a flat or two-dimensional object. Area is measured in square units (square inches, square feet, square centimeters, square meters, etc.). There are several formulas for area on the FCAT Reference Sheet, along with a Key to tell you what each letter in a formula represents. This material is shown below.

| Figure | Name | Formula |
| --- | --- | --- |
| | Triangle | Area (A) = $\frac{1}{2}bh$ |
| | Rectangle | Area (A) = $\ell w$ |
| | Trapezoid | Area (A) = $\frac{1}{2}h(b_1 + b_2)$ |
| | Parallelogram | Area (A) = $bh$ |
| | Circle | Area (A) = $\pi r^2$<br>Circumference = $\pi d = 2\pi r$ |

KEY

$b$ = base; $h$ = height; $\ell$ = length; $w$ = width; $d$ = diameter

Use 3.14 or $\frac{22}{7}$ for $\pi$.

## Rectangles

**Example:** A rectangular room is 4.8 meters wide and 6.6 meters long. What will it cost to carpet this room if carpeting costs $20 per square meter?

Find the area of the room using the formula for area of a rectangle: *Area* = $\ell w$. Substitute the given length and width into the formula: *Area* = $6.6 \times 4.8 = 31.68$.

Multiply the area of the room by $20 to get the cost of carpeting the room: $31.68 \times 20 =$ $633.60.

## Triangles

When using the area formula from the FCAT Reference Sheet, it is helpful to know the parts of a triangle mentioned in the formulas and Key. Important: The height of a triangle is always perpendicular to the base.

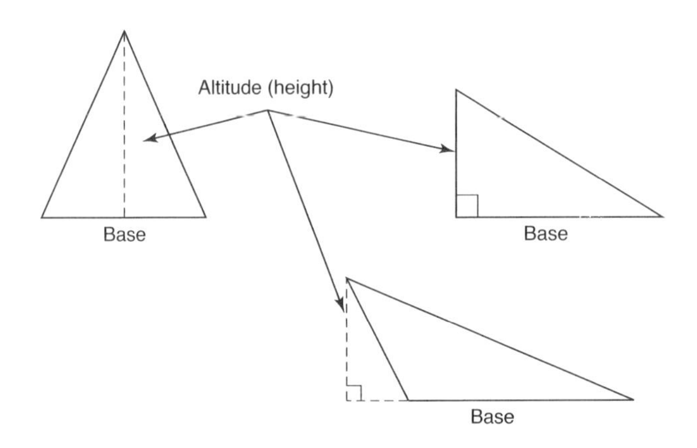


**Example:** The football booster club is making pennants for the fans to carry to the next ball game. The pennants are triangle-shaped.

Part A. Find the area, in square feet, of each pennant.
Part B. Find how many square feet of fabric will be needed to make 50 pennants.

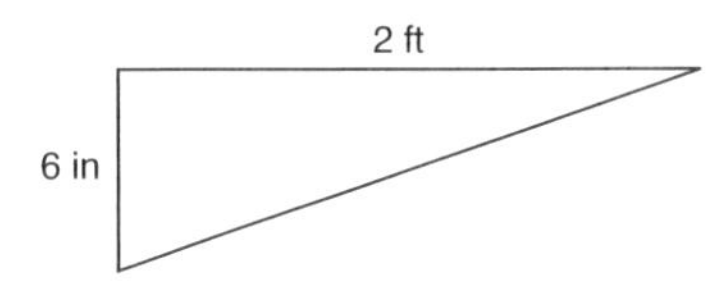


Part A. To find the area in square feet, first make certain all measurements are expressed in feet. *Reminder*: When converting from smaller (inches) to larger (feet), divide (here, by 12): 6 in = 0.5 ft.

Use the formula for area of a triangle: $Area = \frac{1}{2}bh$. Substitute the base (0.5 ft) and the height (2 ft) into the formula: $Area = \frac{1}{2} \times 0.5 \times 2 = 0.5$ sq ft, the amount needed for one pennant.

Part B. Multiply: $0.5 \times 50 = 25$ sq ft are needed for 50 pennants.

## Trapezoids

To use the formula for the area of a trapezoid, it is important to know the parts of a trapezoid. As shown in the drawing, trapezoids have two bases, $b_1$ and $b_2$. The height *(h)* of a trapezoid is perpendicular to both bases.

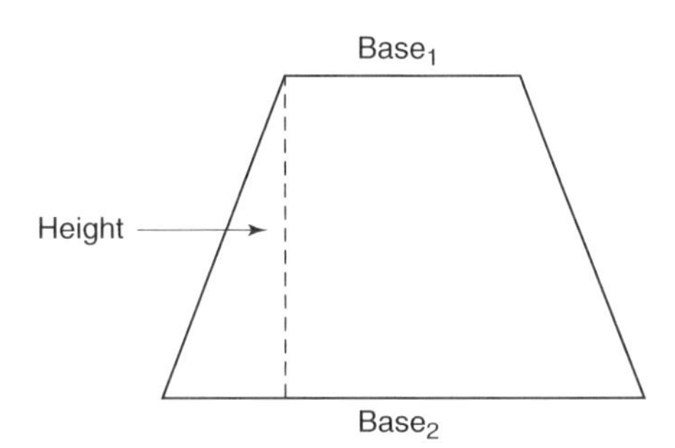


**Example:** Find the area of the trapezoid shown.

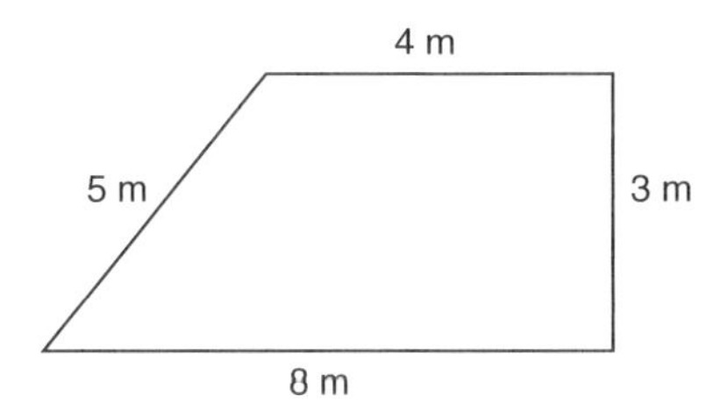


Use the formula for the area of a trapezoid: Area = $\frac{1}{2}h(b_1 + b_2)$. The bases are 4 and 8 m, and the height is 3 m. Substitute values into the formula:

$$\text{Area} = \frac{1}{2} \times 3 \times (4 + 8).$$

First add inside the parentheses: 4 + 8 = 12. Then multiply from left to right: 1.5 × 12 = 18 sq m.

## Parallelograms

The parts of a parallelogram, height and base, are the same as those of a triangle and trapezoid:

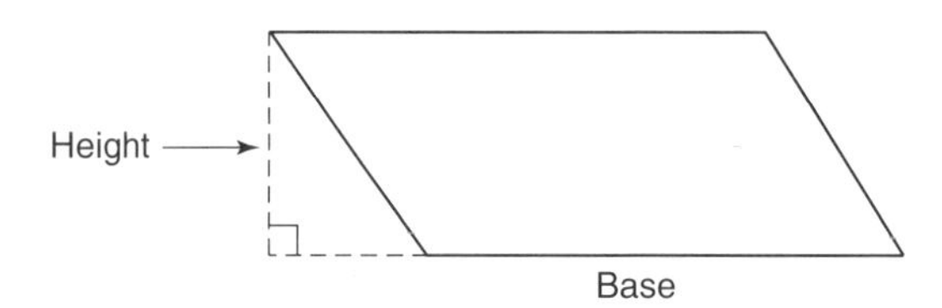


**Example 1:** Find the area of the parallelogram shown.

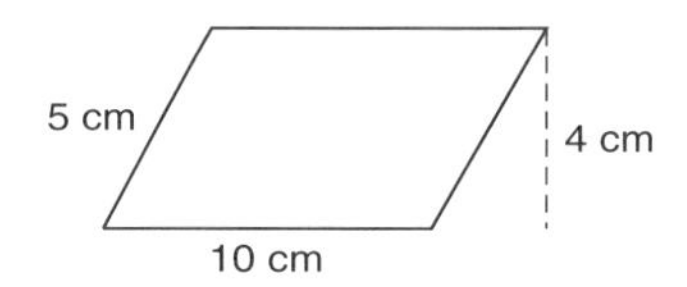


Use the formula Area = $bh$. The base is 10 cm; the height is 4 cm. Substitute these values into the formula:

$$\text{Area} = 10 \times 4 = 40 \text{ sq cm.}$$

**Example 2:** What is the base of the parallelogram shown if the parallelogram has an area of 108 square feet?

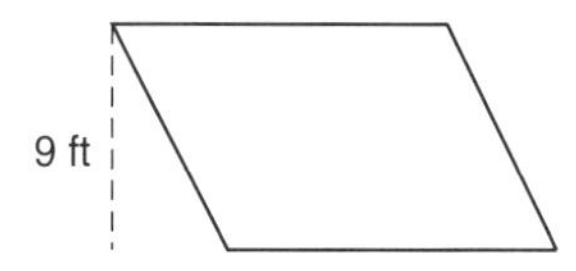


Use the formula Area = $bh$. Substitute the given values: $108 = 9b$. Divide both sides by 9. Base $b = 12$.

## Changes in Dimensions

One of the most difficult types of problems on the FCAT involves changing the dimensions of figures.

**Example:** The area of a square is 25 square feet. If the side lengths of the square are doubled, the area of the new square will be

A. 2 times larger
B. 3 times larger
C. 4 times larger
D. 8 times larger

The correct answer is C; the area will be 4 times larger. Look at the diagram, in which a square with an area of 25 sq ft (5 ft by 5 ft) is placed inside a square with dimensions that are twice as large (10 ft by 10 ft).

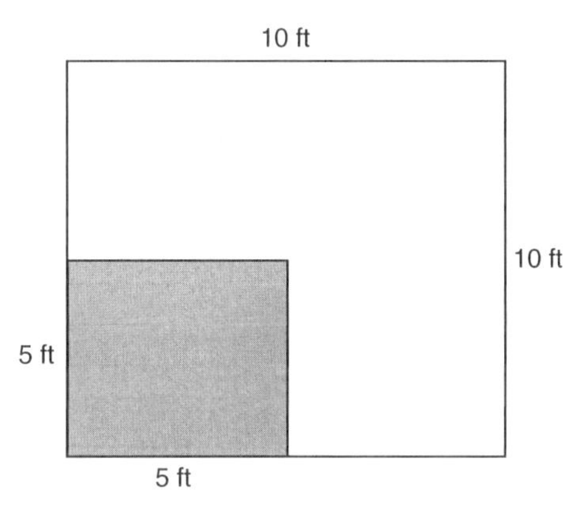


Notice that four of the 5-ft by 5-ft squares will fit inside the new square, which is 10 ft by 10 ft. Although the side lengths are multiplied by 2, the area is 4 times as large, because you multiply the length by 2 and also multiply the width by 2. Therefore, you have actually multiplied by 2 twice.

If you triple the dimensions of the square, or multiply each dimension by 3, the area will be $3 \times 3$ or 9 times larger.

# SAMPLE QUESTIONS

## Perimeters and Areas of Polygons

1. Donny has 100 yards of fencing. He wants to enclose a rectangular garden that will be 40 yards long. How wide, in yards, will the garden be?

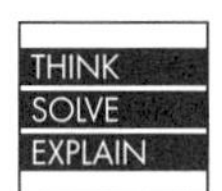


2. Shannon is making an outdoor pen for her peacock. She has a total of 150 feet of fencing to use for three sides of the pen. The fourth side will be a stone wall behind her house.

   Part A. Draw and label a picture of Shannon's peacock pen. Make the largest area possible using whole-number lengths.

   Part B. Explain in words how you determined which dimensions to use to make the pen have the largest area possible.

   ______________________________

   ______________________________

   ______________________________

3. To build endurance, the football team runs laps around the perimeter of the football field shown in the diagram. How many times must the team run around the perimeter of the football field to run about 2 miles? Round your answer to the nearest whole number.

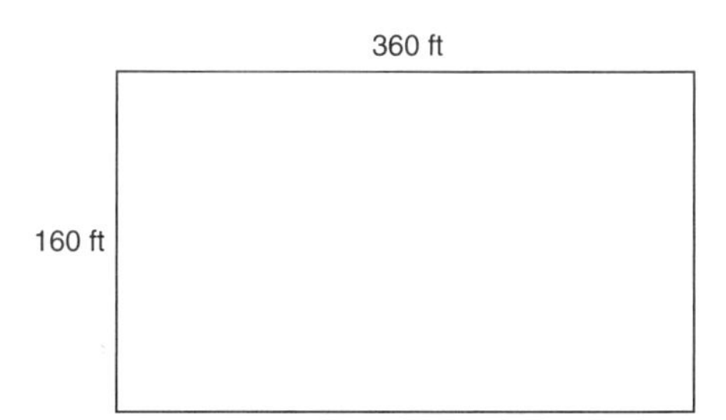


4. Bea has been promoted and is moving from a rectangular office that measures 10 feet by 8 feet to an office that is 3 feet longer and 2 feet wider. How much more area will Bea have in her new office?

   A. 6 sq ft
   B. 50 sq ft
   C. 80 sq ft
   D. 130 sq ft

5. A rectangle measures 20 inches by 10 inches. A square with a side measuring 5 inches is cut out of one corner of the rectangle. How many square inches are there in the remaining piece of the rectangle?

   F. 75
   G. 125
   H. 150
   I. 175

6. A rectangular wall 5 meters long and 2.5 meters high has a rectangular window that is 1 meter wide and 1.75 meters high. What is the area, in square meters, of the wall, not including the window?

7. A 5-inch by 7-inch picture frame holds Angie's favorite photograph. If she has the photograph's dimensions doubled, how many times larger will the area of her new photograph be?

   A. 2      C. 6
   B. 4      D. 8

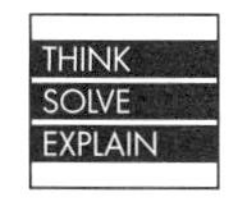


8. Jack and Carol are having their living room tiled and new baseboard put in. The diagram shown is a scale model of their living room.

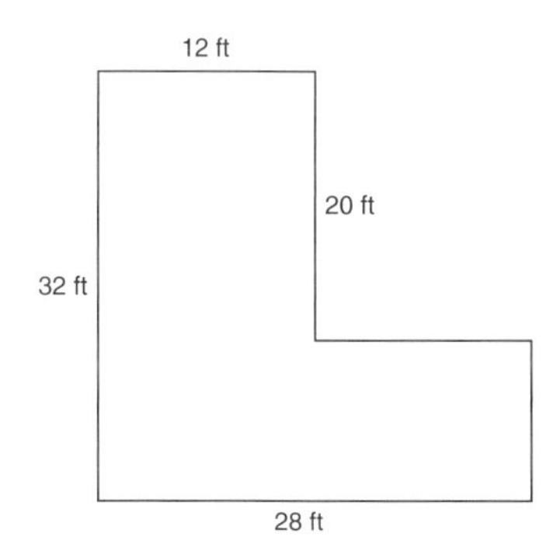


Part A. How many feet of baseboard are needed to go all the way around the living room? If baseboard costs \$0.99 per foot, how much will the baseboard cost? Explain in words, or show your work.

_______________________________

_______________________________

_______________________________

Part B. How many square feet of tile will be needed to cover the entire area? If the tile Jack and Carol selected costs \$6.98 per square foot, how much will it cost to tile the room? Explain in words, or show your work.

_______________________________

_______________________________

_______________________________

9. Find the area of the regular octagon in the diagram if one side measures 10 centimeters.

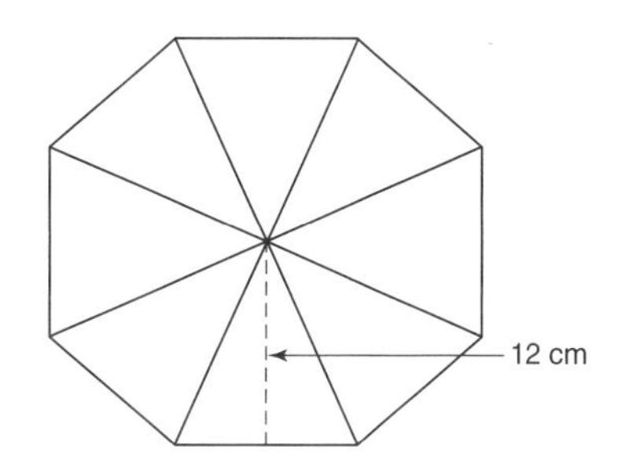


   F. 480 sq cm
   G. 360 sq cm
   H. 120 sq cm
   I. 60 sq cm

10. Jackson is wallpapering his living room. The three walls he needs to cover measure 9 feet by 10 feet. How many square *yards* of wallpaper will he need?

**11.** Find the area of the figure shown in the diagram.

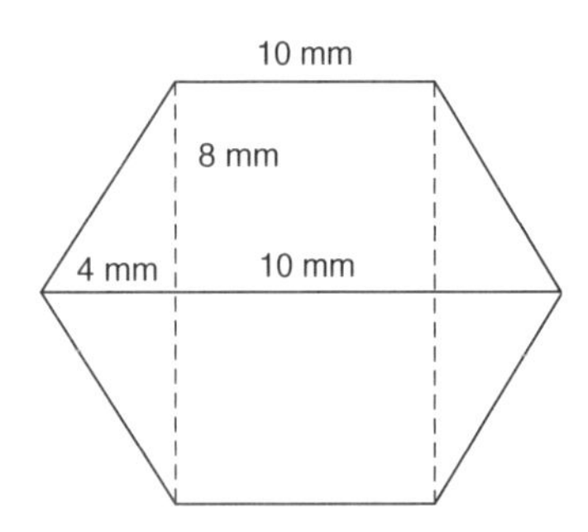


**12.** Find the total number of square feet in the figure shown.

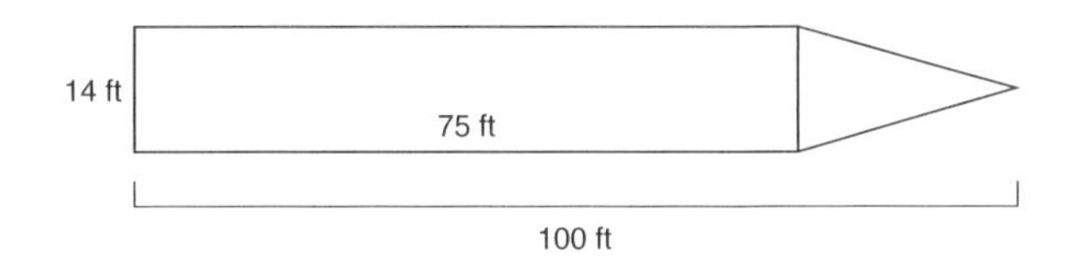


## ANSWERS TO SAMPLE QUESTIONS

1. **10.** Draw and label a diagram like the one shown.

   Subtract: 100 – (40 + 40)

   100 – 80 = 20

   Divide: 20 ÷ 2 = 10

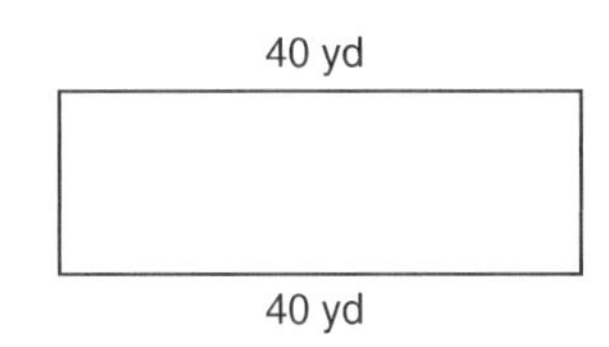


2. Part A.

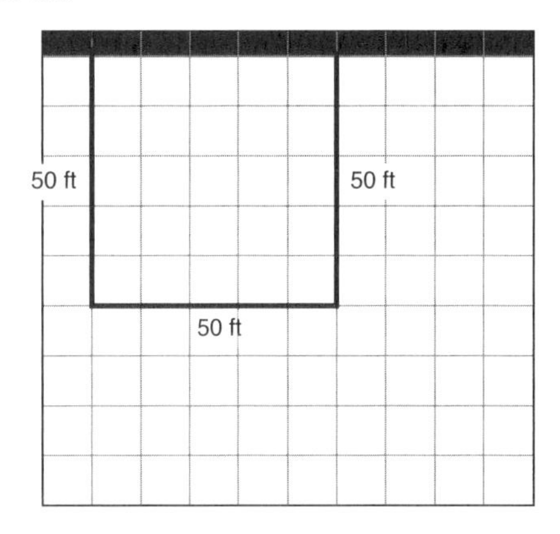


Part B. When squares and rectangles that have the same perimeter (fence) are compared, the square will always have a larger area than the rectangle. Here, the measure of each side can be found by dividing 150 by 3. Each side is 50 ft.

3. **10.** Two miles equals 10,560 ft (5280 × 2). The perimeter of the football field is 2(360) + 2(160) = 1040 ft. 10,560 ÷ 1040 = 10.15, or about 10 times around the field.

4. **B.** The dimensions of the new office are 13 ft long and 10 ft wide. The new area is 130 sq ft. The old office was 10 ft long and 8 ft wide with an area of 80 sq ft. Subtract: 130 – 80 = 50 sq ft.

5. **I.**

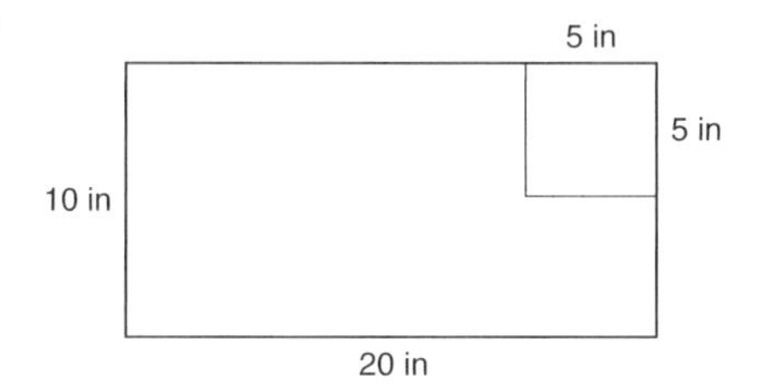


Find the area of the rectangle, and then subtract the area of the square. Area of rectangle: $(20 \times 10)$ minus area of square: $(5 \times 5) = 200 - 25 = 175$.

6. **10.75.**

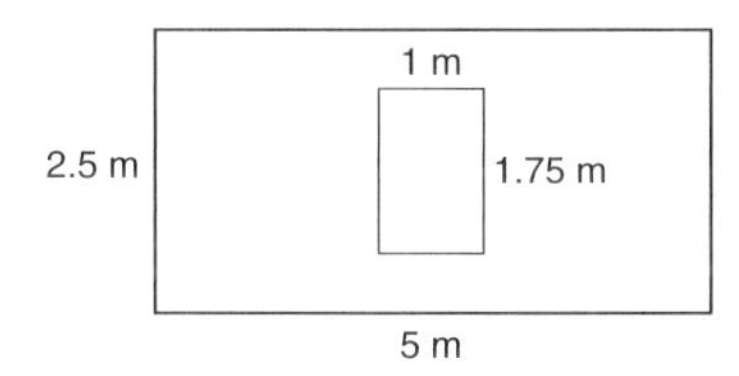


Find the area of the large rectangle, and then subtract the area of the small rectangle. Area of large: $(5 \times 2.5)$ minus area of small: $(1.75 \times 1) = 12.5 - 1.75 = 10.75$.

7. **B.** The old area is 35 sq in. If the dimensions are doubled, the new width $(2 \times 5) = 10$ and the new length $(2 \times 7) = 14$. Then $10 \times 14 = 140$, which is 4 times the original area. *Hint*: Notice that you multiplied by 2 twice.

8. Part A. **120, $118.80.** Find both missing sides: $28 - 12 = 16$ and $32 - 20 = 12$. Add all sides: $12 + 20 + 16 + 12 + 28 + 32 = 120$ ft. To find the cost, multiply: $120 \times 0.99 = \$118.80$ for baseboard.

Part B. **576, $4020.48.** Area of rectangle including the cut-away portion (dotted line in the diagram) is $32 \times 28 = 896$ sq ft. Cut-away portion $(20 \times 16) = 320$. Subtract: $896 - 320 = 576$ sq ft.

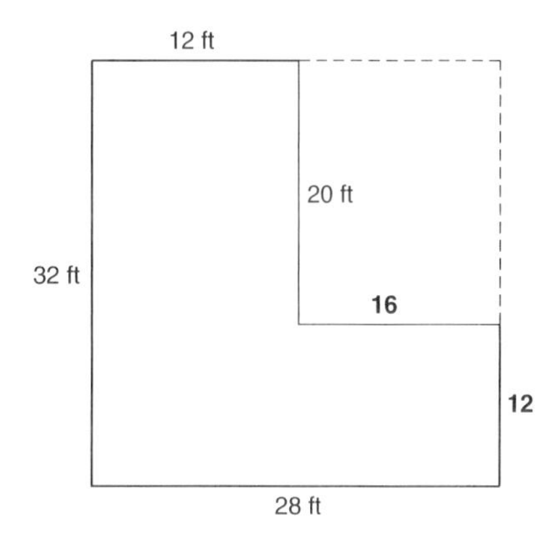


To find the cost, multiply: $576 \times \$6.98 = \$4020.48$.

9. **E.** The octagon has eight triangles. Find the area of one triangle and multiply by 8. Area of one triangle = $\frac{1}{2}bh = 0.5 \times 10 \times 12 =$ 60 sq cm. Area of octagon = $60 \times 8 = 480$ sq cm.

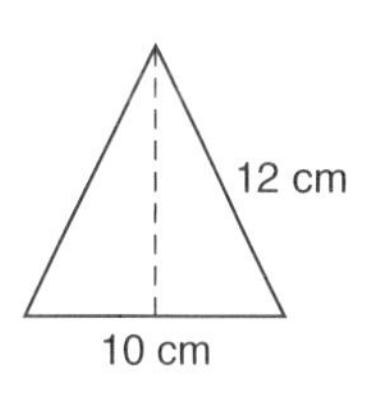

10. **30.** To find total square feet, multiply: $9 \times 10 \times 3 = 270$ sq ft. Convert to square yards by dividing by 9 (there are 9 sq ft in 1 sq yd): $270 \div 9 = 30$.

11. **224 sq mm.** There are two trapezoids in the figure, with $b_1$ as the top base and $b_2$ as the lower base. Label as shown, and use the area formula for trapezoids:

$$\text{Area} = \frac{1}{2}h(b_1 + b_2)$$

$$= \frac{1}{2} \times 8 \times (10 + 18)$$

$$= \frac{1}{2} \times 8 \times 28 = 112 \text{ sq mm.}$$

This is the area of one trapezoid. Multiply by 2 because there are two trapezoids: $112 \times 2 = 224$ sq mm.

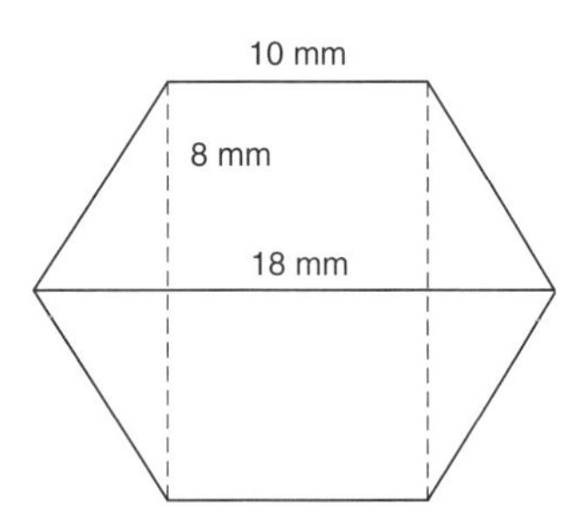


12. **1225.** Find the area of the rectangle and add it to the area of the triangle. The rectangle measures 75 ft by 14 ft. Its area = 1050 sq ft. The triangle has a base of 14 ft and a height of 25 ft (100 ft – 75 ft).

$$\text{Area} = \frac{1}{2} \times 14 \times 25 = 175 \text{ sq ft.}$$

Add the areas: $1050 + 175 = 1225$ sq ft.

# Lesson 12

## Circles

The minute hand of a clock is 6 inches long, as shown in the drawing. How far does the point of the hand travel in 60 minutes?

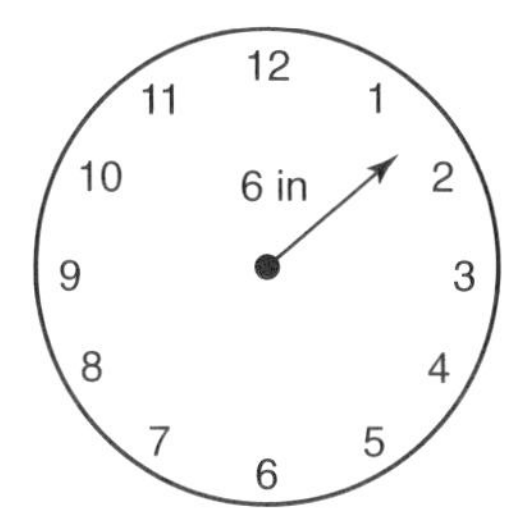


Find the circumference of a circle with a six inch radius using the formula $C = 2\pi r$. Substitute 3.14 for $\pi$ (pi) and 6 for $r$ (radius); then $C = 2 \times 3.14 \times 6 = 37.68$ in.

A *circle* is a set of points of equal distance from the center. A circle is generally named by its center. The circle shown in the diagram is *circle* O.

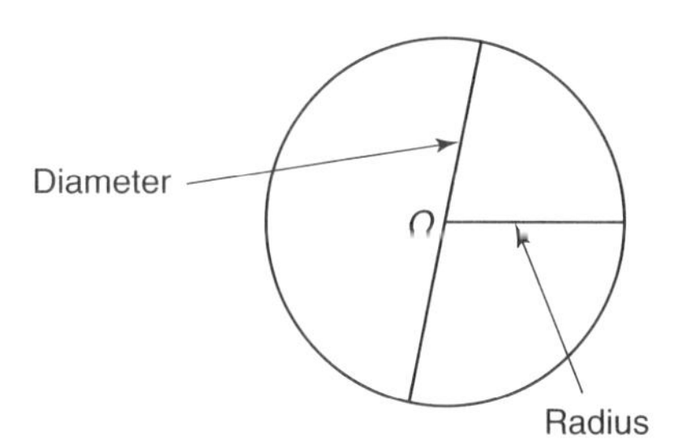


A *radius* is a line segment drawn from the center of the circle to a point on the circle.

A *diameter* is a line segment traveling from one point on the circle, through the center, to another point on the circle. A radius is half the length of a diameter on the same circle.

The perimeter, or distance around the outside edge, of a circle is called the *circumference*. The formula for the circumference of a circle given on the FCAT Reference Sheet is $C = \pi d = 2\pi r$. This is actually two formulas: $C = \pi d$, which uses diameter $d$ multiplied by pi, and $C = 2\pi r$, which uses radius $r$ multiplied by 2. This makes sense because two radii (plural of *radius*) are required to make one diameter. Either 3.14 or $\frac{22}{7}$ is substituted for the Greek letter $\pi$.

The area of a circle measures the space inside the circle. As with polygons, the area of a circle is expressed in square units and is found using the formula $A = \pi r^2$. *Note*: You *must* use the radius $r$ to find the area of a circle.

**Example 1:** Find the circumference and the area of the circle shown in the diagram.

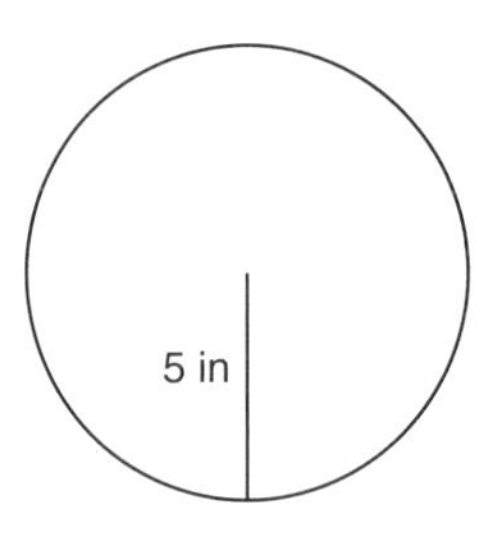


For circumference, use the formula $C = 2\pi r$ because you are given the radius. In the formula, substitute 5 inches for radius and 3.14 for pi.

| | |
|---|---|
| $C = 2\pi r$ | Multiply $2 \cdot \pi \cdot r$. |
| $C = 2 \cdot 3.14 \cdot 5$ | Substitute the values of $\pi$ and $r$. |
| $C = 6.28 \cdot 5$ | |
| $C = 31.4$ in | |

For area, use the formula $A = \pi r^2$. Substitute 5 for radius.

| | |
|---|---|
| $A = \pi r^2$ | Multiply $\pi \cdot r^2$. |
| $A = 3.14 \cdot 5^2$ | Substitute the values of $\pi$ and $r$. |
| $A = 3.14 \cdot 25$ | |
| $A = 78.5$ sq in | |

**Example 2:** Find the circumference and the area of the circle shown in the diagram.

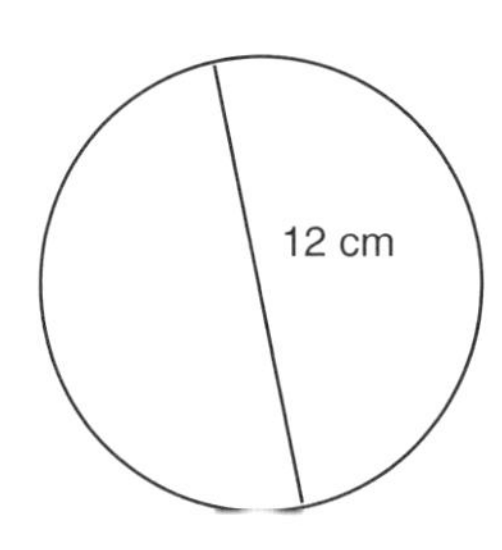


For circumference, use the formula $C = \pi d$ because you are given the diameter. Substitute 12 cm for diameter in the formula and 3.14 for pi.

$C = \pi d$
$C = 3.14 \cdot 12$
$C = 37.68$ cm

For area, use the formula $A = \pi r^2$. Substitute 6 (half the diameter) for radius.

$A = \pi r^2$
$A = 3.14 \cdot 6^2$
$A = 3.14 \cdot 36$
$A = 113.04$ sq cm

Occasionally, you may be asked to find the answer to a problem that uses $\frac{22}{7}$, instead of 3.14, for $\pi$.

**Example 3:** Find the area and the circumference of a circle with a diameter of 14 meters. Use $\frac{22}{7}$ for $\pi$.

To find the area, use the formula $A = \pi r^2$. Substitute 7 (half of 14, the diameter) for the radius.

$A = \pi r^2$

$A = \frac{22}{7} \cdot 7^2$

$A = \frac{22}{7} \cdot 49$

$A = \frac{22}{7} \cdot \frac{49}{1}$ Cross-cancel the 7 and 49.

$A = \frac{22}{1} \cdot \frac{7}{1}$

$A = 154$ sq m

To find the circumference, use the formula $C = \pi d$ because you are given the diameter.

$C = \pi d$

$C = \frac{22}{7} \cdot \frac{14}{1}$ Cross-cancel the 7 and 14.

$C = \frac{22}{1} \cdot \frac{2}{1} = 44$ m

If you have the circumference or the area, you may need to work backward to find the radius or diameter.

**Example 1:** The space shuttle's orbit is roughly circular. On each orbit the shuttle travels 12,874 miles. What is the radius of this circle?

Use the circumference formula with radius to solve this problem (you are given a distance around a circular object). Substitute 12,874 for circumference C, and solve the equation.

$C = 2\pi r$

$12{,}874 = 2 \cdot 3.14 \cdot r$

$12{,}874 = 6.28r$

$2050 = r$ Divide both sides by 6.28.

The radius is 2050 mi.

**Example 2:** A revolving sprinker sprays water in all directions. It covers 1500 square feet with water as it turns. If you stand 25 feet away from the sprinkler, will you get wet? Why or why not?

Use the area formula to solve this problem because you are given "square feet." Substitute 1500 for area $A$, and solve the equation.

$A = \pi r^2$

$1500 = 3.14r^2$

$477.7 = r^2$ Divide both sides by 3.14.

$\sqrt{477.7} = \sqrt{r^2}$ Because $r$ is "squared," take the square root to solve.

$21.9 = r$ Water sprays out 21.9 ft. If you are 25 ft away, you will not get wet.

# SAMPLE QUESTIONS

## Circles

1. Find the area and the circumference of the circle shown in the diagram.

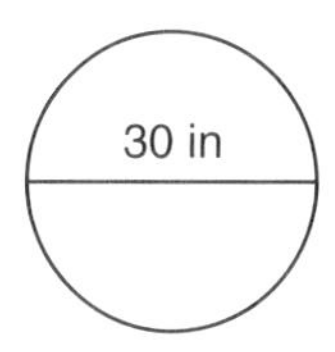


A. $A$ = 706.5 sq in; $C$ = 188.4 in
B. $A$ = 706.5 sq in; $C$ = 94.2 in
C. $A$ = 2826 sq in; $C$ = 188.4 in
D. $A$ = 2826 sq in; $C$ = 94.2 in

2. A tanker truck leaks ammonia gas out in all directions for a distance of 7 miles. How many square miles are affected by the gas? Use $\frac{22}{7}$ for $\pi$.

3. The wheels on Jennifer's bicycle have a radius of 10 inches. How far does the bicycle travel with each turn of the wheels?

F. 62.8 in
G. 125.6 in
H. 120 in
I. 251.2 in

4. Brandon ran 8 laps around a circular track with a radius of 110 feet. ESTIMATE how far he ran.

A. 660 ft
B. 1320 ft
C. 2640 ft
D. 1 mi

5. Find the area, in square inches, of the circle inside the square in the diagram.

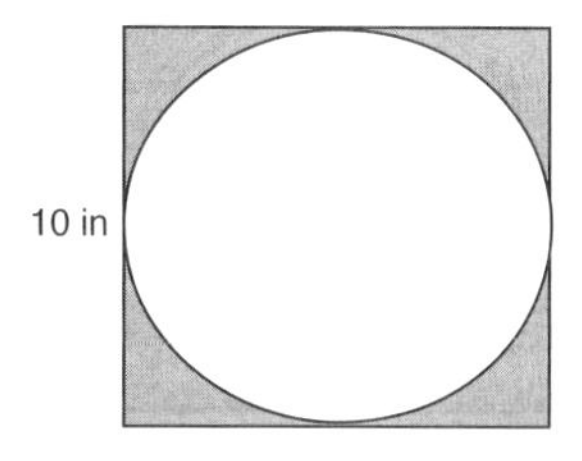


6. Using the diagram in Problem 5, find the area, in square inches, of the shaded portion.

7. A child is lost in the woods. The searchers have decided that the child could not have traveled more than 2 miles in any direction since he became lost. How many square miles must they search to find the child?

F. 6.28
G. 25.12
H. 12.56
I. 50.24

8. Angie's mother is making her a circular poodle skirt for the 50's party at her school. The skirt is made of felt, and Angie wants it to be 30 inches long. How many square feet of felt will Angie's mother need to purchase to make the skirt? Round your answer to the nearest whole number.

9. Find the area of the shaded region in the diagram. Use $\frac{22}{7}$ for $\pi$. Round to the tenths place.

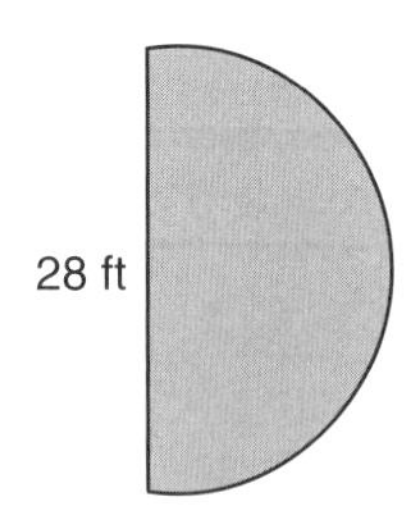


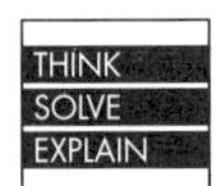


10. The diagram shows a race track. The radius to the outside lane is 40 yards. The radius to the inside lane is 30 yards.

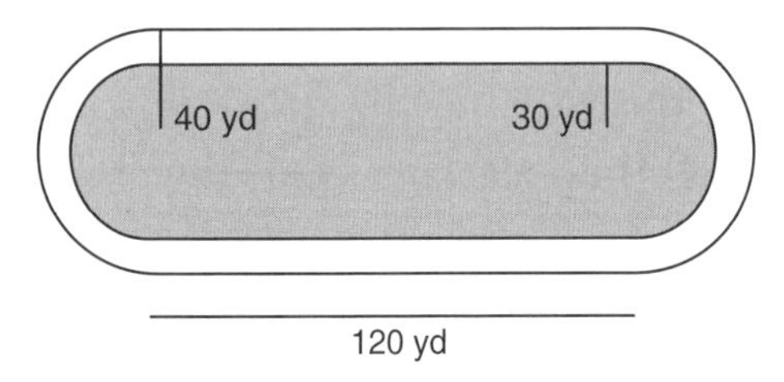


Part A. How much of a head start would a runner on the outside lane need? Show how you arrived at your answer.

______________________________

______________________________

______________________________

Part B. If the shaded area inside the track is to be planted in grass, how many square yards of grass will be needed?

______________________________

______________________________

______________________________

# ANSWERS TO SAMPLE QUESTIONS

1. **B.** Area = $3.14 \cdot 15^2 =$ $3.14 \cdot 225 = 706.5$ sq in. Circumference = $3.14 \cdot 30 =$ 94.2 in.

2. **154.** Because the question asks for *square* miles, you will need to find the area of the circle. Think of the tanker truck as being at the center of a circle. The radius of the circle will be 7 miles. Use $A = \pi r^2$, and substitute $\frac{22}{7}$ for $\pi$ and 7 for $r$:

   $A = \frac{22}{7} \cdot 7^2 = \frac{22}{7} \cdot \frac{49}{1} = \frac{22}{1} \cdot \frac{7}{1}$
   $= 154.$

3. **F.** The question asks for the distance traveled, so use the circumference formula: $C = 2\pi r$. $C = 2 \cdot 3.14 \cdot 10 = 62.8$ in.

4. **D.** When estimating, round first. Round 3.14 to 3. Use the circumference formula: $C = 2\pi r$ and multiply by 8. $C = 2 \cdot 3 \cdot 110 \cdot 8 = 5280$ ft = 1 mi.

5. **78.5.** The circle has a radius of 5 in. $A = 3.14 \cdot 5^2 = 3.14 \cdot 25 = 78.5$.

6. **21.5.** Find the area of the square and subtract from it the area of the circle (found in problem 5). Area of square = $10 \cdot 10 = 100$; $100 - 78.5 = 21.5$.

7. **H.** The question asks for *square* miles. Use the area formula: $A = 3.14 \cdot 2^2 = 3.14 \cdot 4 = 12.56$.

8. **20.** Because the problem asks for *square* feet, first convert 30 into feet by dividing by 12: 30 in = 2.5 ft. Then use the area formula: $A = 3.14 \cdot 2.5^2 = 3.14 \cdot 6.25 = 19.625$. Round to 20.

9. **308 sq ft.** The figure represents half a circle. Find the area of a circle with a radius of 14 ft, and divide by 2.
   $A = \pi r^2 \div 2$
   $A = \frac{22}{7} \cdot \frac{14^2}{1} \div 2$
   $A = \frac{22}{7} \cdot \frac{196}{1} \div 2$
   $A = \frac{4312}{7} \div 2 = 616 \div 2$
   $A = 308$

10. Part A. **62.8 yd.** Distance around outside lane = circumference of larger circle with radius of 40 added to 120 two times:

    $2 \cdot 3.14 \cdot 40 + 120 + 120 =$
    $251.2 + 120 + 120 = 491.2$.

    Distance around inside lane = circumference of smaller circle with radius of 30 added to 120 two times:

    $2 \cdot 3.14 \cdot 30 + 120 + 120 =$
    $188.8 + 120 + 120 = 428.8$.

    Subtract: $491.2 - 428.8 = 62.8$ yd head start.

    Part B. **10,026 sq yd.** Find the area of the smaller circle, using a radius of 30, and add to the area of the rectangle measuring 60 by 120.

    Area of circle = $3.14 \cdot 30^2 =$
    $3.14 \cdot 900 = 2826$ sq yd
    Area of rectangle = $60 \cdot 120 = 7200$ sq yd
    Add: $2826 + 7200 = 10{,}026$ sq yd

# Lesson 13

## Surface Area and Volume

A cube measures 8 centimeters on an edge. What is the total surface area of the cube?

All six faces of a cube are the same size. Find the area of one face, and multiply by 6.

Area of one face = $8 \cdot 8 = 64$. Multiply by 6: $64 \cdot 6 = 384$ sq cm.

A *rectangular solid* is a three-dimensional figure made up of six rectangles or squares. Each of these is called a *face*. Faces are joined together at an *edge*, and edges are joined together at a *vertex*. The base of the rectangular solid can be the top or bottom of the figure.

A cereal box is an example of a rectangular solid. Cubes are also rectangular solids. The three dimensions are length, width, and height.

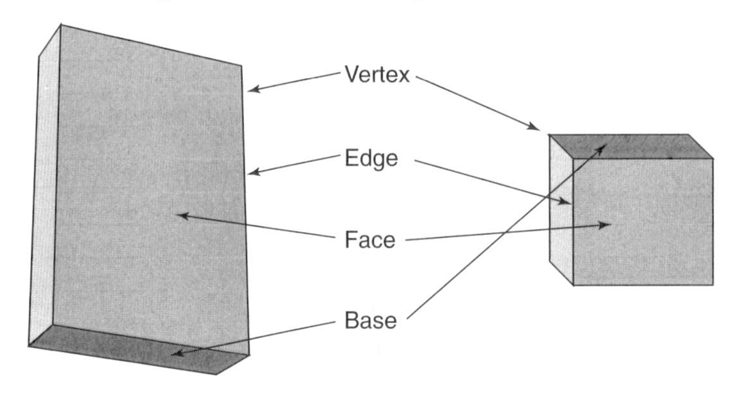


### SURFACE AREA OF A RECTANGULAR PRISM

*Surface area* is the sum of the areas of all six faces of a rectangular solid or prism and is expressed in square units. Simply find the area of each face and add. If a cereal box were taken apart and flattened, you could see the six surfaces more clearly, as shown in the diagram.

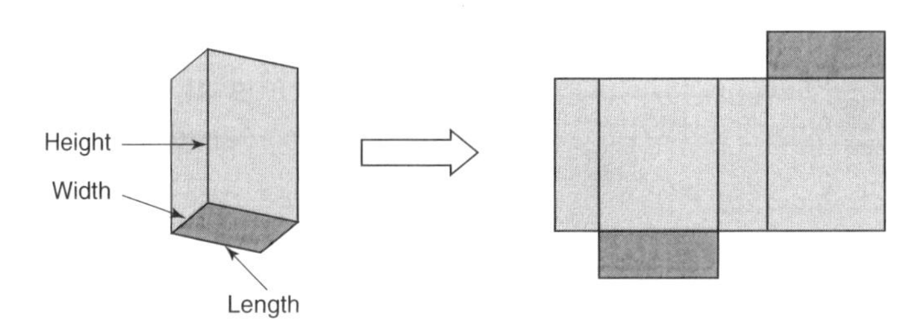

**Example 1:** How many square inches of cardboard are needed to make a cereal box that is 8 inches long, 3 inches wide, and 11 inches high?

Use the formula for surface area of a rectangular solid: Total surface area (SA) = $2(\ell w) + 2(hw) + 2(\ell h)$.

Substitute 8 for $\ell$, 3 for $w$, and 11 for $h$:

$SA = 2(\ell w) + 2(hw) + 2(\ell h)$
$SA = 2 \cdot (8 \cdot 3) + 2 \cdot (11 \cdot 3) + 2 \cdot (8 \cdot 11)$. Perform all multiplications first.
$SA = 48 + 66 + 176 = 290$ sq in

**Example 2:** If the total surface area of a cube is 864 square millimeters, what is the area of one face?

Since the total surface area of a cube equals 6 times the area of one side, set up the following equation: $864 = 6A$, where $A$ equals the area of one face.

Divide both sides by 6: $\frac{864}{6} = \frac{6A}{6}$. Then $144 = A$.

The area of one face is 144 sq mm.

**Example 3:** Cube $A$ has an edge length of 3 millimeters and cube $B$ has an edge length of 6 millimeters. What is the ratio of the surface area of cube $A$ to cube $B$?

Remember from Lesson 12 that areas are expressed in square units. The area of one face of cube $A$ is $3 \cdot 3 = 9\ \text{mm}^2$ and the area of one face of cube $B$ is $6 \cdot 6 = 36\ \text{mm}^2$. The ratio of the areas of one face in the two cubes is the same as the ratio of the areas of all faces.

The area of cube $A$ is to the area of cube $B$ as 9 is to 36. Express this relationship as a fraction:

$\frac{9}{36} = \frac{1}{4}$. The ratio is 1 to 4.

## VOLUME OF A RECTANGULAR PRISM

Volume measures the amount of space *inside* an object or the amount that it will hold, called the *capacity*. Volume can be expressed in many units: ounces, liters, milliliters, quarts, gallons, or cubic units such as cubic inches.

The formula on the FCAT Reference Sheet for the volume of a rectangular prism is

$$V = \ell wh\text{: Volume} = \text{length} \times \text{width} \times \text{height}.$$

**Example 1:** Find the volume of a rectangular solid with length 10 feet, width 5 feet, and height 4 feet, as shown in the diagram.

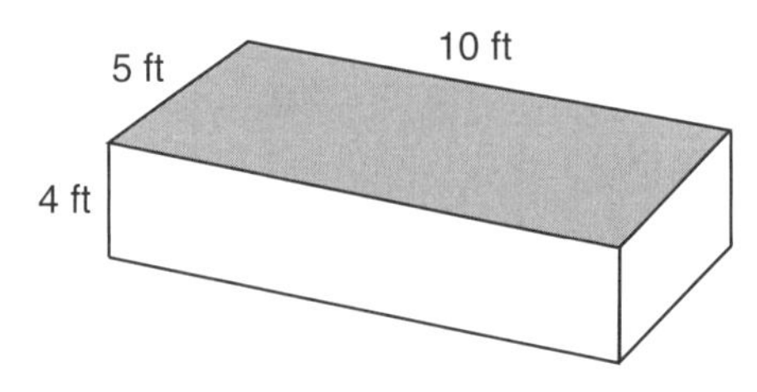


Use the formula $V = \ell wh$, substitute the values for length, width, and height, and multiply:

$$V = 10 \cdot 5 \cdot 4 = 200 \text{ cu ft.}$$

**Example 2:** The volume of a cube is given as 1000 cubic inches. Which of the following could be the length of one side?

A. 500 in
B. 333 in
C. 100 in
D. 10 in

One method of solving this problem is to work backward, using the possible answers to find the volume. Use each answer, and multiply by itself 3 times to see which one gives 1000: $500 \cdot 500 \cdot 500$ is too large, as is $333 \cdot 333 \cdot 333$, and $100 \cdot 100 \cdot 100$. Only $10 \cdot 10 \cdot 10$ gives the correct answer. Therefore, D is correct.

**Example 3:** A pond that measures 20 meters by 10 meters holds 500 cubic meters of water. How deep is the pond?

Use the formula $V = \ell wh$, and substitute the given values. You know that 500 represents the volume because it is expressed in cubic meters. Substitute 20 m for length and 10 m for width.

$V = \ell wh$
$500 = 20 \cdot 10 \cdot h$
$500 = 200h$
$2.5 = h$ Divide both sides by 200.

The pond is 2.5 m deep.

**Example 4:** Use the rectangular prism shown in the diagram to answer the questions.

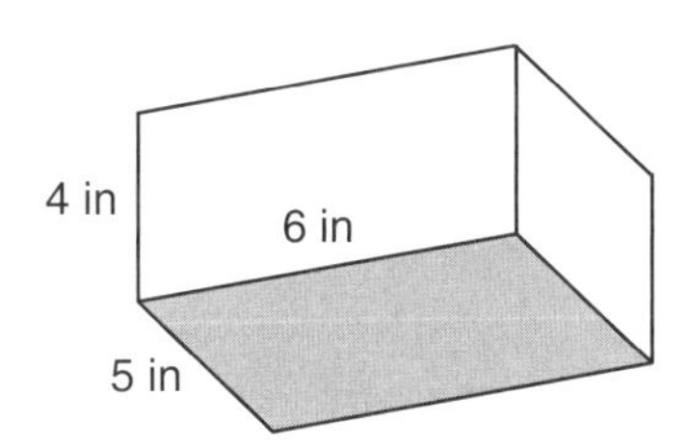


Part A. If the length is doubled, how many times larger will the volume be?

Part B. If the length and width are doubled, how many times larger will the volume be?

Part C. If the length, width, and height are all doubled, how many times larger will the volume be?

Part A. If only the length is doubled, the volume will be two times as large because you multiplied by 2 only one time.

Old volume: $V = \ell \cdot w \cdot h$
$V = 6 \cdot 5 \cdot 4$
$V = 120$ cu in

New volume: $V = 2\ell \cdot w \cdot h$
$V = 2 \cdot 6 \cdot 5 \cdot 4$
$V = 240$ cu in

Part B: If the length and width are doubled, the volume will be four times as large because you multiplied by 2 two times.

Old volume: $V = \ell \cdot w \cdot h$
$V = 6 \cdot 5 \cdot 4$
$V = 120$ cu in

New volume: $V = 2\ell \cdot 2w \cdot h$
$V = 2 \cdot 6 \cdot 2 \cdot 5 \cdot 4$
$V = 480$ cu in

Part C: If the length, width and height are all doubled, the volume will be eight times as large because you multiplied by 2 three times.

Old volume: $V = \ell \cdot w \cdot h$
$V = 6 \cdot 5 \cdot 4$
$V = 120$ cu in

New volume: $V = 2\ell \cdot 2w \cdot 2h$
$V = 2 \cdot 6 \cdot 2 \cdot 5 \cdot 2 \cdot 4$
$V = 960$ cu in

Another way to find the volume of a rectangular prism is to find the area of the base $B$ and multiply by the height $h$. This formula is written as $V = Bh$.

**Example 5:** Find the volume of a rectangular prism with a base area of 29 square feet and a height of 15 feet.

The volume of a rectangular prism can be found by multiplying the area of the base by the height: $V = 29 \cdot 15 = 435$ cu ft.

## SURFACE AREA OF A CYLINDER

To find the surface area of a cylinder such as a can, add the areas of the top and bottom of the cylinder to the area of the label portion, called the *lateral side*.

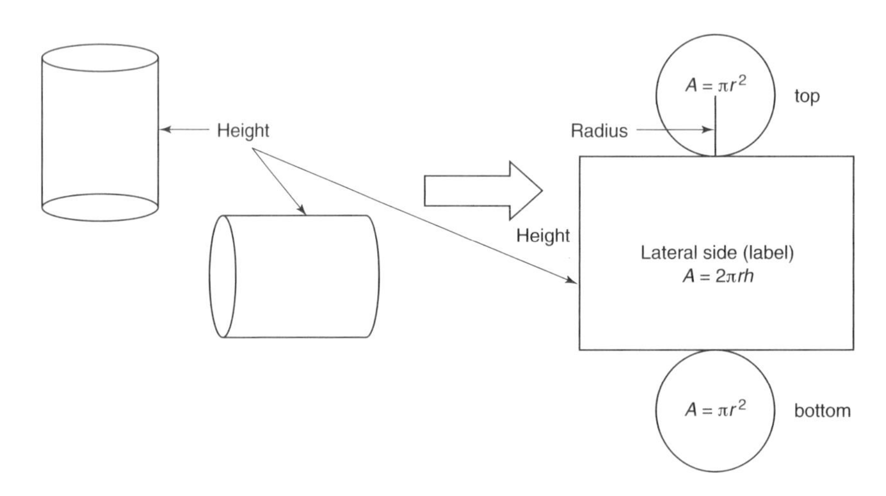


The formula for the surface area of a cylinder (also on the FCAT Reference Sheet) is a combination of the formulas shown above: SA = areas of the two circles, $2\pi r^2$, + area of the lateral side, $2\pi rh$.

**Example 1:** Find the surface area of a cylinder with a radius of 5 inches and a height of 12 inches. Use 3.14 for $\pi$.

Use the formula $SA = 2\pi r^2 + 2\pi rh$, and substitute the given values.

$SA = 2\pi r^2 + 2\pi rh$
$SA = 2 \cdot 3.14 \cdot 5^2 + 2 \cdot 3.14 \cdot 5 \cdot 12$
$SA = 6.28 \cdot 25 + 6.28 \cdot 5 \cdot 12$
$SA = 157 + 376.8 = 533.8$ sq in

**Example 2:** How many square inches of paper are needed to make a label for a can 7 inches tall with a radius of 3 inches? Use $\frac{22}{7}$ for $\pi$.

The formula for the surface area of a cylinder is $SA = 2\pi r^2 + 2\pi rh$. Remove the part of the formula that represents the areas of the top and bottom: $2\pi r^2$. The part that remains is the area of the label portion, or the lateral area.

$SA = 2\pi rh$

$SA = 2 \cdot \frac{22}{7} \cdot 3 \cdot 7$ Substitute given values.

$SA = \frac{44}{7} \cdot 3 \cdot 7$

$SA = \frac{132}{7} \cdot 7 = 132$ sq in

## VOLUME OF A CYLINDER

The volume of a cylinder is measured in cubic units or in other volume units such as ounces, gallons, liters, and milliliters. The volume measures how much the cylinder will hold when it is filled.

The formula for the volume of a cylinder is given on the FCAT Reference Sheet: $V = \pi r^2 h$. It is the area formula for a circle (the base) multiplied by the height of the cylinder.

**Example 1:** Find the volume of the cylinder shown in the diagram to the nearest cubic centimeter.

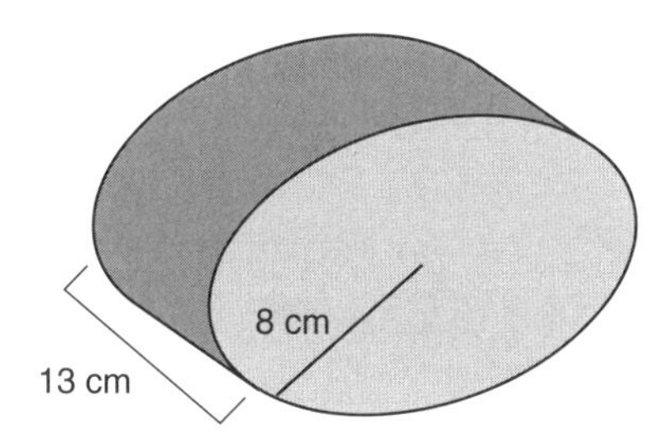


The height of the cylinder is 13 cm, and its radius is 8 cm. Use the formula $V = \pi r^2 h$, and substitute the given radius and height into the formula. Use 3.14 for $\pi$.

$V = \pi r^2 h$
$V = 3.14 \cdot 8^2 \cdot 13$
$V = 3.14 \cdot 64 \cdot 13$
$V = 2612.48$ cu cm

**Example 2:** Find the volume of a can of soup $2\frac{3}{4}$ inches in diameter and 4 inches tall.

Divide $2\frac{3}{4}$ by 2 to find the radius.

$\frac{11}{4} \div \frac{2}{1} = \frac{11}{4} \times \frac{1}{2} = \frac{11}{8}$. Radius $= \frac{11}{8} = 1.375$ in

Use the formula $V = \pi r^2 h$.

$V = \pi r^2 h$
$V = 3.14 \cdot 1.375^2 \cdot 4$
$V = 3.14 \cdot 1.89 \cdot 4 = 23.7$ cu in

**Example 3:** A steel bar has a radius of $\frac{1}{2}$ inch and is 36 inches long. If the radius is doubled, how many times more steel will be needed to make the bar?

Use the formula $V = \pi r^2 h$. When the radius is doubled, it goes from $r$ to $2r$. The formula uses $r^2$. If the radius is doubled, you must use $(2r)^2$, which equals $4r^2$. Therefore, four times more steel is needed to make the bar. Check by using the formula and substituting the values for radius and length.

Old volume:
$V = \pi r^2 h$
$V = 3.14 \cdot 0.5^2 \cdot 36$
$V = 3.14 \cdot 0.25 \cdot 36$
$V = 28.26$ cu in

New volume:
$V = \pi(2r)^2 h$
$V = 3.14 \cdot (2 \cdot 0.5)^2 \cdot 36$
$V = 3.14 \cdot 1^2 \cdot 36$
$V = 113.04$ cu in, which is 4 times 28.26 cu in

If the radius and the length of the steel bar in Example 3 are both doubled, how many times more steel will be needed to make the bar?

When the radius is doubled, the bar requires 4 times more steel. If the length is doubled too, the bar requires $4 \cdot 2 = 8$ times more steel.

**Example 4:** A cylinder has a volume of 486.4 cubic millimeters and a radius of 8 millimeters. What is the height of the cylinder?

Use the formula $V = \pi r^2 h$, and substitute the given values.

$V = \pi r^2 h$
$V = 3.14 \cdot 8^2 \cdot h$
$486.4 = 3.14 \cdot 64 \cdot h$
$486.4 = 200.96 \cdot h$
$2.4 = h$ Divide both sides by 200.96.

The height of the cylinder is 2.4 mm.

**Example 5:** Find the volume of a cylinder with a base area of 42.5 square centimeters and a height of 9 centimeters.

The volume of a cylinder can be found by multiplying the base area by the height: $42.5 \cdot 9 = 382.5$ cu cm.

# SAMPLE QUESTIONS

## Surface Area and Volume

1. The poured granite top for a coffee table is made in the shape of a rectangular solid. It has a volume of 8 cubic feet. The customer wants an end table with dimensions that are exactly half the dimensions of the coffee table. How many times less should he have to pay for the top to the end table?

   A. 16     C. 4
   B. 8     D. 2

2. A metal band is wrapped around a barrel 24 inches in diameter and 4 feet high. How long, in inches, should the metal band be to go around the barrel with no overlap.

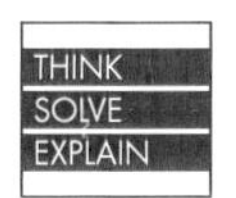


3. A cubic centimeter made of wood weighs about 3.2 grams. Andi's teacher showed the class a wooden cube with an edge length of 5 centimeters.

   Part A. How many cubic centimeters are in the teacher's cube? Show your work, or explain in words how you arrived at your answer.

   ______________________________

   ______________________________

   Part B. How much does the teacher's cube weigh?

   ______________________________

   ______________________________

4. Dan gave Jessica a birthday gift with ribbon wrapped all the way around the box with no overlaps. How many inches of ribbon did he use?

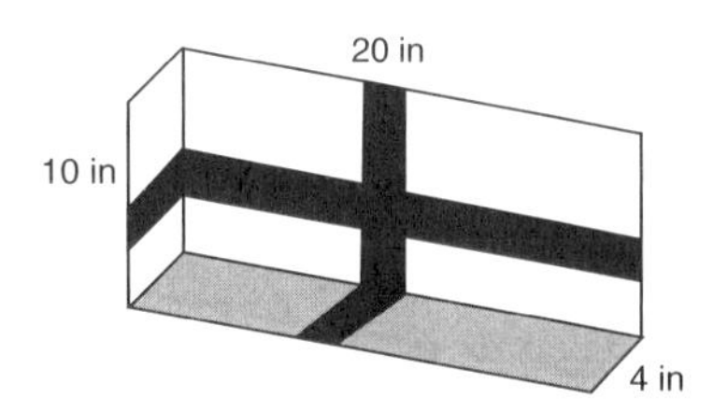


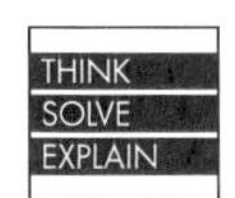


5. A water trough is in the shape of a trapezoidal prism, as shown in the drawing.

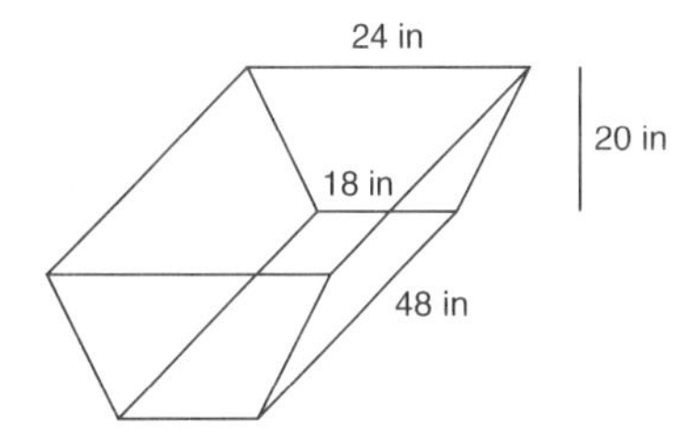


   Part A. If the volume of the trough can be found using this formula: Volume = area of base times height, what is the volume of the water trough in cubic inches? Show your work, or explain in words how you arrived at your answer.

   ______________________________

   ______________________________

   ______________________________

Part B. If 1 cubic foot = 7.5 gallons, how many gallons will the trough hold?

______________________________

______________________________

______________________________

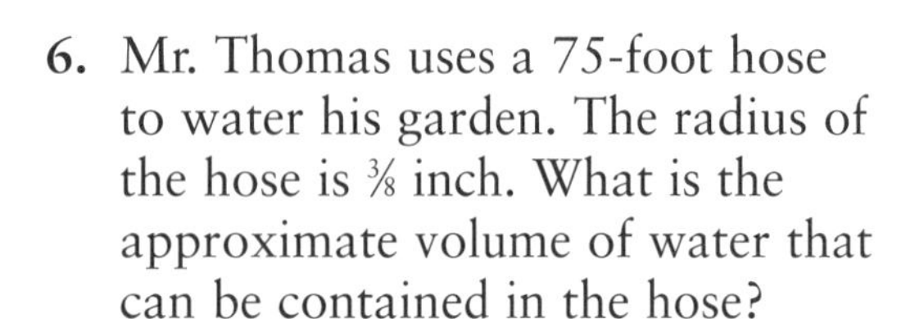

6. Mr. Thomas uses a 75-foot hose to water his garden. The radius of the hose is $\frac{3}{8}$ inch. What is the approximate volume of water that can be contained in the hose?

   F. 1600 cu in  H. 107 cu in
   G. 400 cu in  I. 56 cu in

7. Maryellen purchased an aquarium as shown in the diagram. She filled it with water until the water level was 5 centimeters from the top. How many cubic centimeters of water are in the aquarium?

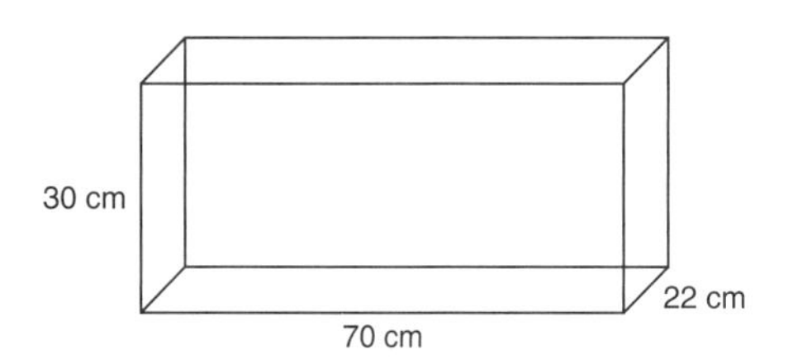


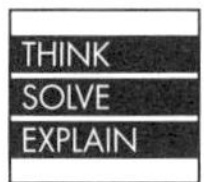


8. A rectangular prism is 8 units long, 4 units wide, and 2 units high.

   Part A. Double the dimensions of the prism, and find its surface area.

   ______________________________

   ______________________________

   ______________________________

   Part B. Explain in words how the surface area of the prism changes when the length, width, and height are doubled.

   ______________________________

   ______________________________

   ______________________________

9. Concrete has a mass of about 2500 kilograms per cubic meter. About how many kilograms of concrete will be needed to build a driveway 13 meters long, 18 meters wide, and 9 centimeters thick?

## ANSWERS TO SAMPLE QUESTIONS

1. **B.** Each dimension is multiplied by one-half:
$\frac{1}{2} \times \frac{1}{2} \times \frac{1}{2} = \frac{1}{8}$.
The cost of the end table should be $\frac{1}{8}$ of the cost of the coffee table top, or 8 times less.

2. **75.36.** The distance around the barrel is circumference. The height of the barrel is unnecessary information. Use $C = \pi d$:
$C = 3.14 \times 24 = 75.36$.

3. Part A. There are $5 \times 5 \times 5 =$ **125** cu cm in the teacher's cube.

   Part B. The teacher's cube weighs $125 \times 3.2 =$ **400 g.**

4. **76.** Add:
$20 + 4 + 20 + 4 + 10 + 4 + 10 + 4$.

5. Part A. **20,160.** The base is a trapezoid. The formula for the area of a trapezoid is $A = \frac{1}{2}h(b_1 + b_2)$.
Substitute given values:
$A = \frac{1}{2} \cdot 20 \cdot (24 + 18)$.
Area of base $= 10 \cdot 42 = 420$ sq. in.
Multiply area of base by height (length of trough): $420 \cdot 48 = 20{,}160$ cu in.

   Part B. **87.5.** There are $12 \cdot 12 \cdot 12$ cu inches in 1 cu ft.
Divide: $20{,}160 \div 1728 = 11.67$ cu ft.
Multiply: $11.67 \cdot 7.5 = 87.5$ gal in the trough.

6. **G.** A garden hose is a cylinder. First convert 75 ft to inches: 75 by 12 = 900 in. This represents the "height" of the cylinder. The radius of the hose is $\frac{3}{8}$, or 0.375 in. Use the formula for volume of a cylinder:
$V = 3.14 \cdot (0.375)^2 \cdot 900 =$
$3.14 \cdot 0.140625 \cdot 900 =$
397.406 cu in.

7. **38,500.** Multiply: $25 \cdot 70 \cdot 22 =$ 38,500.

8. Part A. The surface area is **448 square units.** If the dimensions are doubled, the new dimensions will be 16 units long, 8 units wide, and 4 units high. Substitute the new dimensions into the surface area formula for rectangular prisms:
$SA = 2(\ell w) + 2(wh) + 2(\ell h)$.
$SA = 2(16 \cdot 8) + 2(8 \cdot 4) + 2(16 \cdot 4) =$
$256 + 64 + 128 = 448$ square units.

   Part B. Doubling the dimensions causes the surface area to be **4 times larger** than it originally was.

9. **52,650.** Change 9 cm to meters by dividing by 100. Then find the volume by multiplying length times width times height:
$13 \cdot 18 \cdot 0.09 = 21.06$ cu m of concrete.
Multiply by 2500 kg per cu m:
$21.06 \cdot 2500 = 52{,}650$ kg.

# Lesson 14

## Congruent and Similar Polygons

### CONGRUENT POLYGONS

When two or more figures have the same size and shape, they are *congruent*.

The symbol for congruency is $\cong$. To indicate that the triangles shown are identical, we use the congruency statement: $\triangle ABC \cong \triangle DEF$, which states that triangle $ABC$ is congruent to triangle $DEF$.

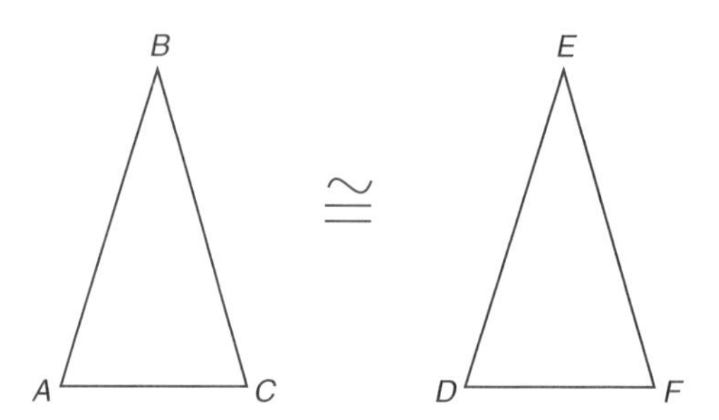


In the figure above, the vertices are labeled so that similar parts correspond. $A$ corresponds with $D$, $B$ corresponds with $E$, and $C$ corresponds with $F$. Each angle and each side in triangle $ABC$ must be congruent with the corresponding angle and side in triangle $DEF$. Side $\overline{AB}$ corresponds with $\overline{DE}$, $\overline{AC}$ corresponds with $\overline{DF}$, and $\overline{BC}$ corresponds with $\overline{EF}$.

In the figures below, the markings on $\triangle LMN$ correspond with the markings on $\triangle PQR$. Therefore, all angles and sides are congruent. When all angles and sides are congruent, the figures are congruent, even though they are oriented differently.

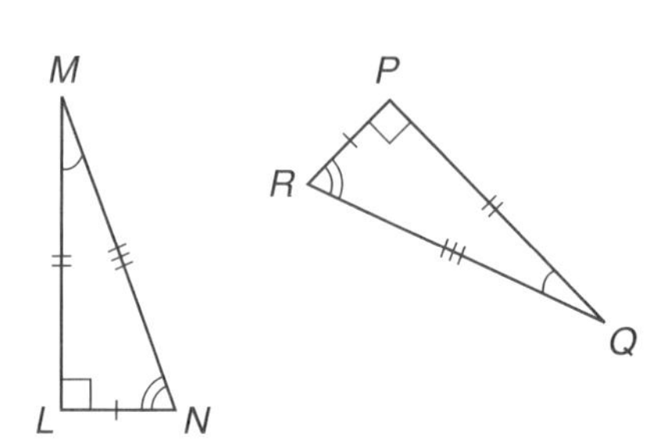

**Example:** The two quadrilaterals shown are congruent.

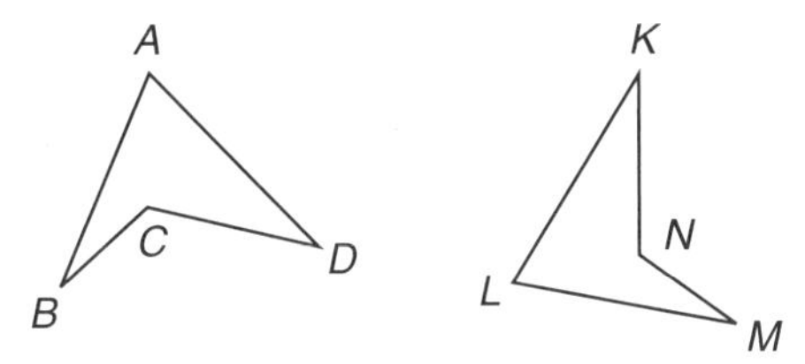


A. Which vertex corresponds with $K$? $D$ corresponds with $K$.
B. Which vertex corresponds with $L$? $A$ corresponds with $L$.
C. Which side corresponds with $\overline{BC}$? $\overline{MN}$ corresponds with $\overline{BC}$.
D. Which side corresponds with $\overline{LK}$? $\overline{AD}$ corresponds with $\overline{LK}$.

## SIMILAR POLYGONS

Similar polygons, such as the triangles shown, have the same shape, but are not the same size. The symbol for similarity is ~. We say that $\triangle ABC \sim \triangle DEF$.

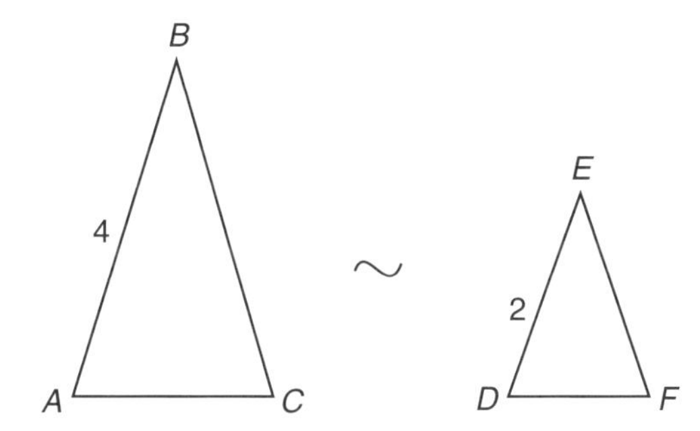


In similar polygons, corresponding angles are congruent while corresponding sides have the same ratio or scale factor. Because corresponding sides have the same ratio, the figures are proportional.

The triangles shown above are similar. Corresponding angles are congruent (have the same measure). The scale factor is $\frac{1}{2}$, meaning that the sides of the smaller triangle are half the length of the sides of the larger triangle.

Because the two triangles below are similar, their sides are proportional. We can set up ratios of *corresponding sides*:

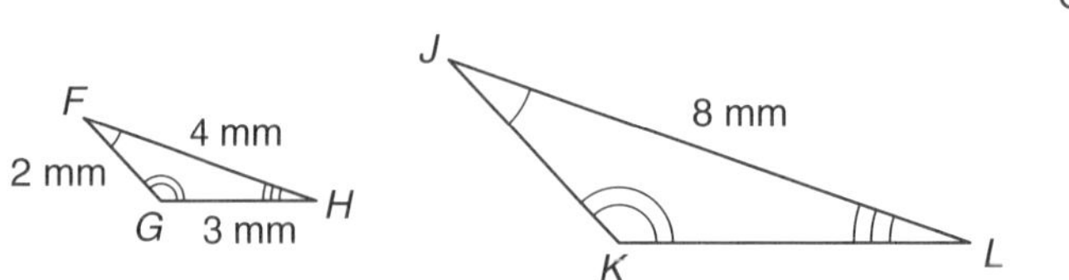


Corresponding angles:

$\angle F \cong \angle J$
$\angle G \cong \angle K$
$\angle H \cong \angle L$

$$\frac{4}{8} = \frac{3}{\overline{KL}} = \frac{2}{\overline{JK}}$$

By using two of these ratios to form a proportion, it is possible to find the lengths of sides $KL$ and $JK$.

$\frac{4}{8} = \frac{3}{\overline{KL}}$ Multiply 8 by 3, and divide by 4: $\overline{KL} = 6$.

$\frac{4}{8} = \frac{2}{\overline{JK}}$ Multiply 8 by 2, and divide by 4: $\overline{JK} = 4$.

# SAMPLE QUESTIONS

## Congruent and Similar Polygons

1. The right triangles shown are congruent: $\triangle ABC \cong \triangle DEF$.

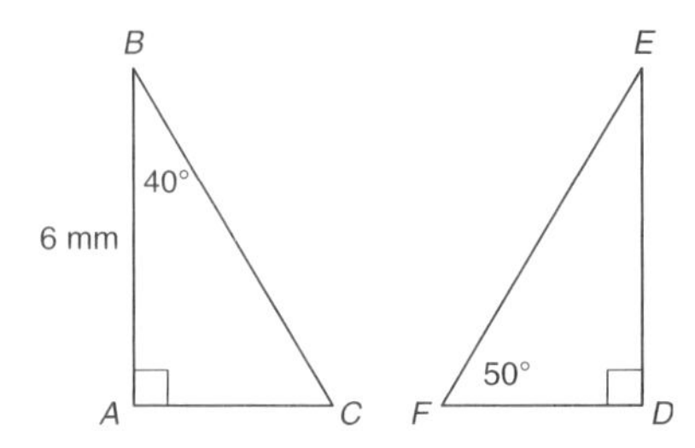


Which statement is NOT true?

A. $\angle C = 50°$  C. $\overline{DE} = 6$ mm
B. $\angle E = 40°$  D. $\overline{FE} = 8$ mm

2. The two triangles in the figure are similar. How many degrees are in $\angle L$?

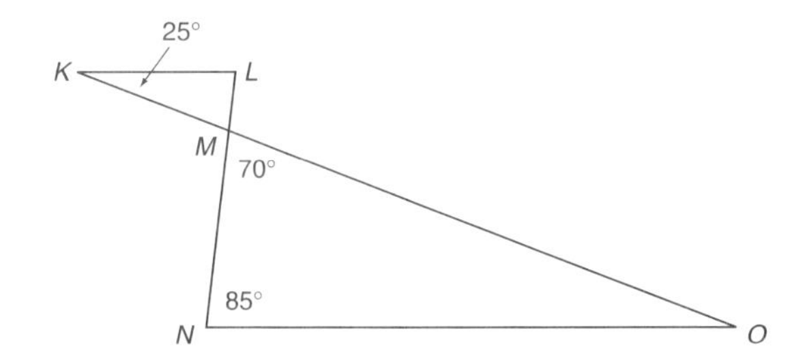


3. If $\triangle QRS \sim \triangle TUV$, which of the following correctly completes the statement $\frac{\overline{QR}}{\overline{TU}} = \frac{\overline{QS}}{?}$

F. $\overline{UV}$  H. $\overline{TV}$
G. $\overline{ST}$  I. $\overline{RS}$

4. A manufacturer makes a clock in two sizes. The clocks are identical except for their sizes. The hour hand of the smaller clock is 3 inches long, and the minute hand is 4 inches long. How long, in inches, is the hour hand of the larger clock if the minute hand is 12 inches long?

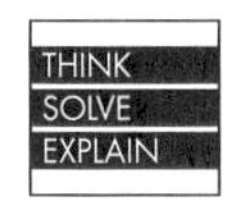


5. Are the two figures shown on the grid similar? Why or why not? Explain in words, and use mathematical terms in your answer.

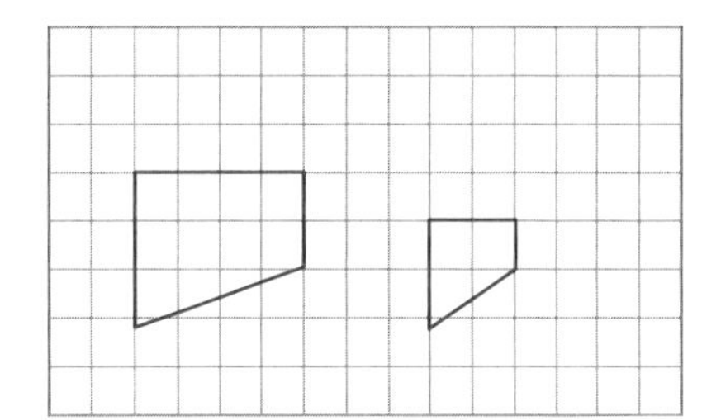

________________________________

________________________________

________________________________

6. In the diagram, triangle $ABD$ is similar to triangle $ECD$. Which proportion can be used to find the length of $\overline{DA}$?

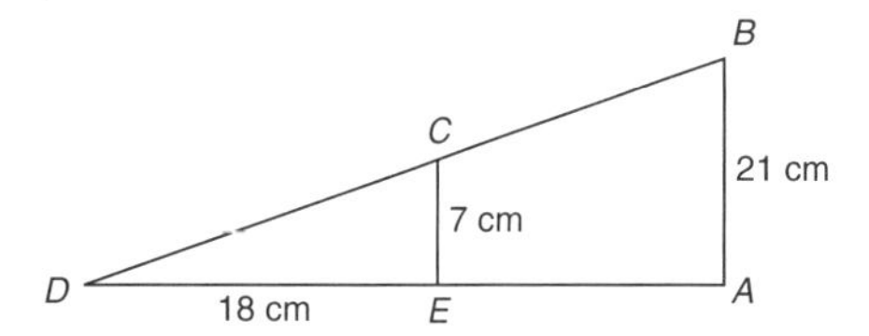


A. $\frac{21}{18} = \frac{7}{\overline{DA}}$

C. $\frac{21}{7} = \frac{18}{\overline{DA}}$

B. $\frac{21}{7} = \frac{\overline{DA}}{18}$

D. $\frac{7}{18} = \frac{\overline{DA}}{18}$

7. The polygons shown below are congruent. Vertex $A$ corresponds with which vertex of the other polygon?

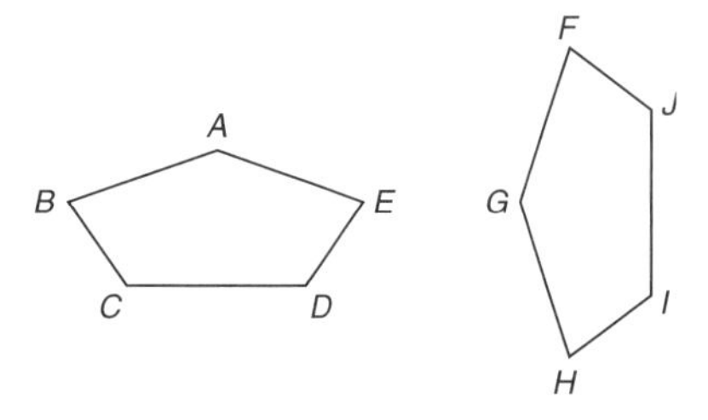


F. $F$

G. $G$

H. $H$

I. $I$

8. If the *scale factor* of two similar figures is a ratio of the corresponding sides, what is the scale factor of the figures shown?

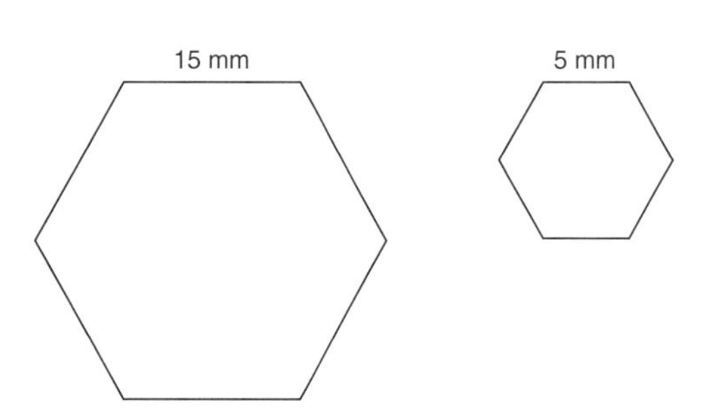


9. What is the scale factor of two circles if the first circle has a radius of 2 and the second circle has a radius of 8?

A. 4

C. $\frac{1}{4}$

B. 2

D. $\frac{1}{2}$

10. Recall from Lessons 12 and 13 that area scale factors are squared (second power) and volume scale factors are cubed (third power). Use what you learned from these lessons to fill in the missing values in the table.

| Scale Factor | Area Scale Factor | Volume Scale Factor |
|---|---|---|
| $\frac{1}{2}$ | | |
| $\frac{1}{3}$ | | |
| | $\frac{4}{9}$ | |
| $\frac{1}{4}$ | | |
| $\frac{2}{5}$ | | |
| | $\frac{9}{16}$ | |

# ANSWERS TO SAMPLE QUESTIONS

1. **D.** Not enough information is given to suggest that $\overline{FE}$ = 8 mm.

2. **85.** Angles *L* and *N* are corresponding angles in similar triangles. Therefore, they are congruent and have the same measure.

3. **H.** The symbol ~ indicates that the two triangles are similar. *Q* corresponds with *T*, *R* corresponds with *U*, and *S* corresponds with *V*. Therefore, $\overline{QR}$ corresponds with $\overline{TU}$, and $\overline{QS}$ corresponds with $\overline{TV}$.

4. **9.** This question can be answered using a proportion that compares the lengths of the hour and minute hands of the smaller clock to the lengths of the hour and minute hands of the larger clock:
$\frac{\text{hour}}{\text{minute}} = \frac{\text{hour}}{\text{minute}}$.
Substitute the actual measurements:
$\frac{3}{4} = \frac{x}{12}$. Solve by multiplying 3 by 12 and then dividing by 4.

5. Ratios of corresponding sides are 4:2, 3:2, and 2:1. In similar figures the ratios would be equal. The figures are not similar because the ratios are not equal and proportional.

6. **B.** It is easier to see the proportional relationship if the triangles are drawn separately, as shown. Side $\overline{BA}$ (21 cm) corresponds with side $\overline{CE}$ (7 cm), and side $\overline{DA}$ corresponds with side $\overline{DE}$ (18 cm).

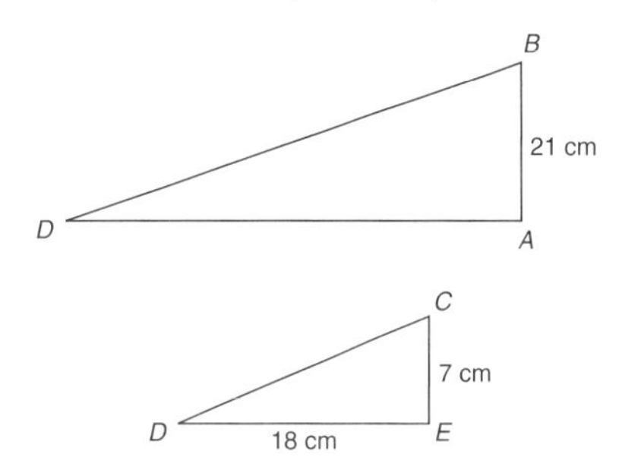


7. G.

8. $\frac{1}{3}$. Find the scale factor by dividing the side length of the second figure (5) by the side length of the first figure (15):
$\frac{5}{15} = \frac{1}{3}$. The sides of the smaller figure are one-third the length of the sides of the larger figure.

9. **A.** Divide: 8 ÷ 2 = 4.

10.

| Scale Factor | Area Scale Factor | Volume Scale Factor |
|---|---|---|
| $\frac{1}{2}$ | $\frac{1^2}{2^2} = \frac{1}{4}$ | $\frac{1^3}{2^3} = \frac{1}{8}$ |
| $\frac{1}{3}$ | $\frac{1^2}{3^2} = \frac{1}{9}$ | $\frac{1^3}{3^3} = \frac{1}{27}$ |
| $\frac{2}{3}$ | $\frac{2^2}{3^2} = \frac{4}{9}$ | $\frac{2^3}{3^3} = \frac{8}{27}$ |
| $\frac{1}{4}$ | $\frac{1^2}{4^2} = \frac{1}{16}$ | $\frac{1^3}{4^3} = \frac{1}{64}$ |
| $\frac{2}{5}$ | $\frac{2^2}{5^2} = \frac{4}{25}$ | $\frac{2^3}{5^3} = \frac{8}{125}$ |
| $\frac{3}{4}$ | $\frac{3^2}{4^2} = \frac{9}{16}$ | $\frac{3^3}{4^3} = \frac{27}{64}$ |

# Lesson 15

## Coordinate Geometry—Distance

Give the coordinates of the points labeled on the coordinate grid.

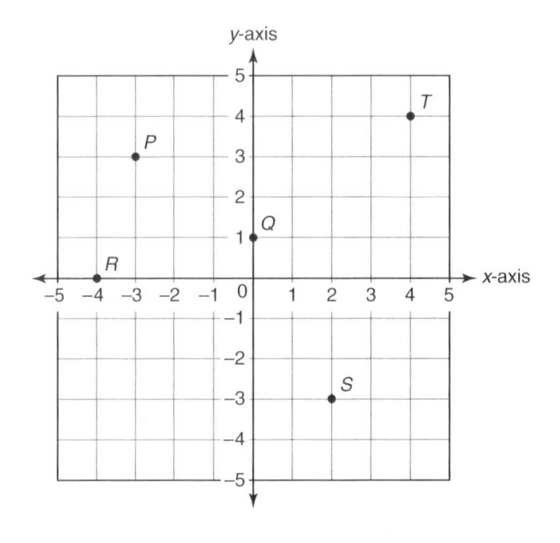


$P(-3, 3)$; $Q(0, 1)$; $R(-4, 0)$; $S(2, -3)$; $T(4, 4)$.

The term *coordinate geometry* refers to the relationships between points and lines on a coordinate plane. The coordinate plane or grid shown below is divided by an *x-axis* and a *y-axis* into four parts called *quadrants*. The two axes cross at the *origin*.

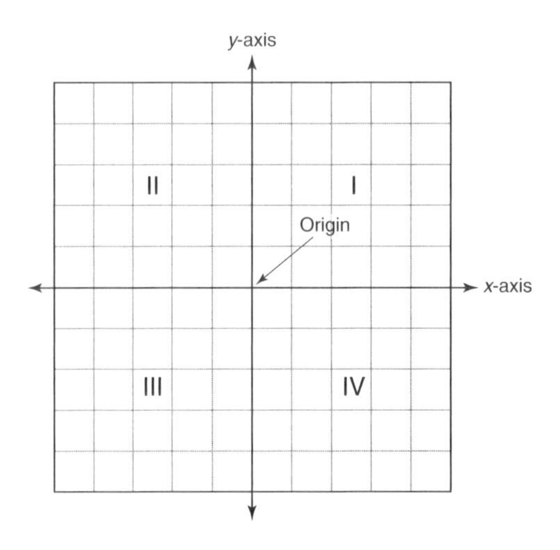

Every point graphed on the grid requires an *x-coordinate*, which identifies the horizontal distance from the origin, and a *y-coordinate*, which identifies the vertical distance from the origin. The two coordinates are written as an *ordered pair (x, y)*.

**Example:** Note where each point on the grid is plotted:
$A(4, -2)$, $B(0, 3)$, $C(3, 3)$, $D(-2, 4)$, $E(-5, -5)$, $F(-1, 0)$

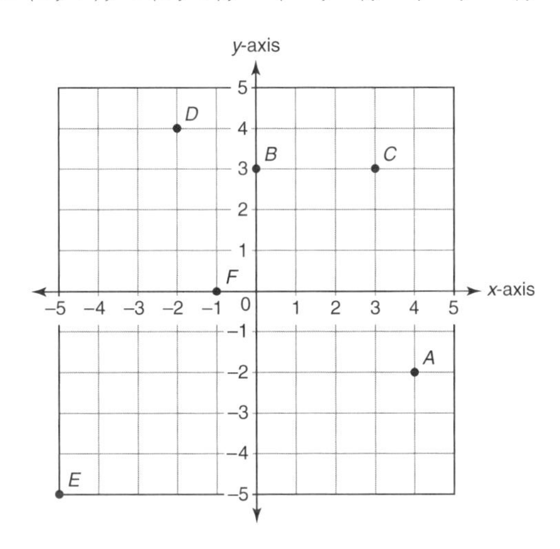


## MIDPOINT OF A LINE

The midpoint of a line segment is the point that is exactly halfway between the endpoints.

**Example 1:** Find the midpoint of $\overline{QR}$.

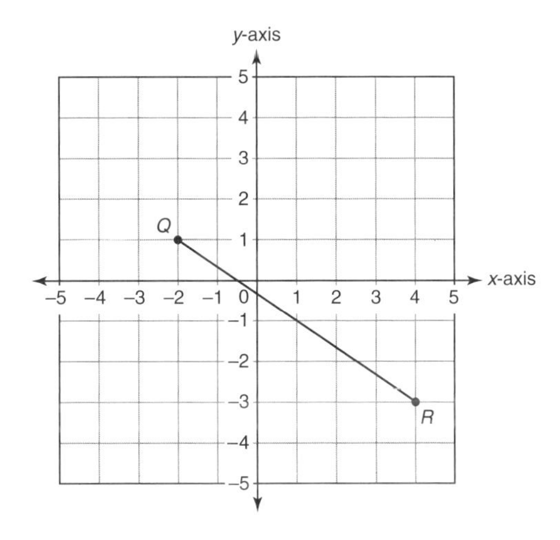


The midpoint of $\overline{QR}$ is located at $(1, -1)$.

**Example 2:** Find the midpoint of $\overline{CD}$.

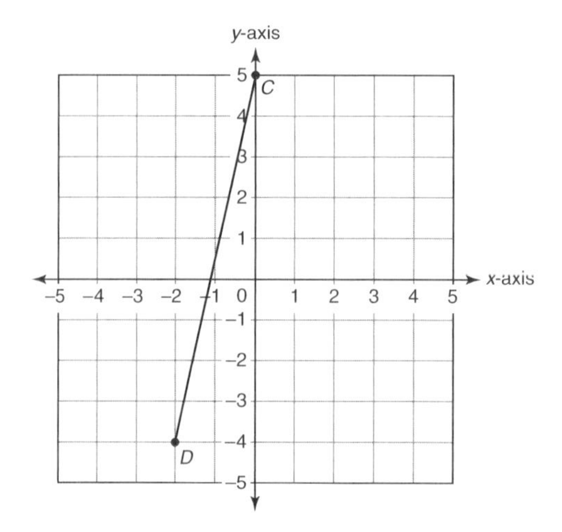


The midpoint of $\overline{CD}$ can be found by using the formula $\frac{(x_1 + x_2)}{2}, \frac{(y_1 + y_2)}{2}$.
The formula tells you to add the two $x$-coordinates and divide by 2, then add the two $y$-coordinates and divide by 2. Use $C(0, 5)$ and $D(-2, -4)$:

$$\frac{[0+(-2)]}{2}, \frac{[5+(-4)]}{2} = \left(\frac{-2}{2}, \frac{1}{2}\right) = (-1, 0.5).$$

The midpoint is located at (–1, 0.5).

## DISTANCE

If two points are on the same horizontal line, you can find the distance between them by subtracting their $x$-coordinates. If the points are given to you but are not graphed, you can tell they are on the same horizontal line if their $y$-coordinates are the same.

**Example 1:** Find the distance between the points shown on the graph.

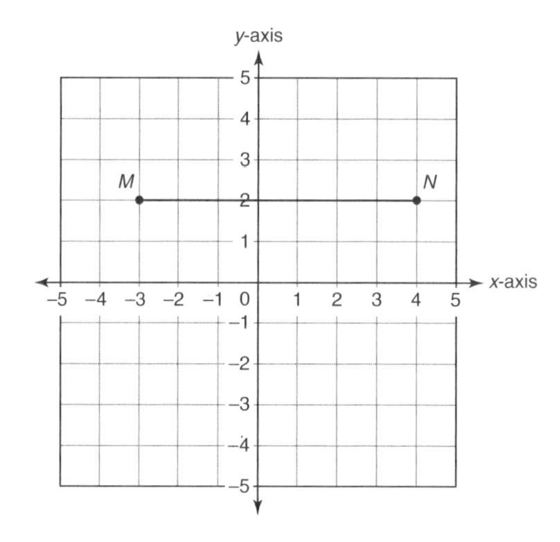

The points shown are $M(-3, 2)$ and $N(4, 2)$. These points have the same $y$-coordinate; therefore they are on the same horizontal line. To find their distance, subtract the $x$-coordinates: $4 - (-3) = 7$. The points are 7 spaces apart.

**Example 2:** Find the distance between the points $(-5, 4)$ and $(3, 4)$.

These points have the same $y$-coordinate; therefore they are on the same horizontal line. Subtract the $x$-coordinates: $3 - (-5) = 8$. The points are 8 spaces apart.

When two points are on the same vertical line, you can find the distance between them by subtracting their $y$-coordinates. If the points are given but are not graphed, you can tell they are on the same vertical line if their $x$-coordinates are the same.

**Example 1:** Find the distance between the points shown on the graph.

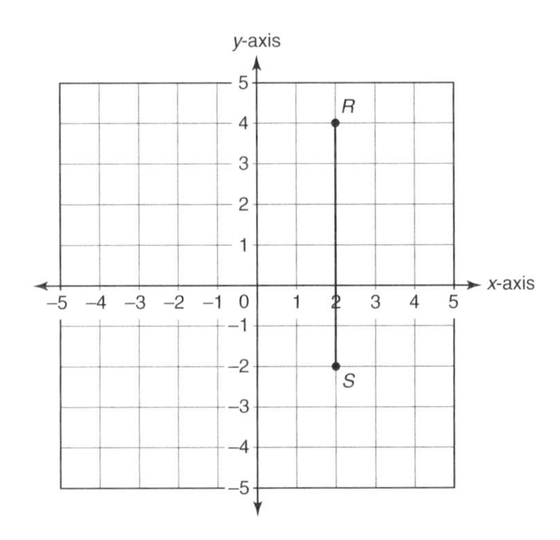


The points shown are $R(2, 4)$ and $S(2, -2)$. They have the same $x$-coordinate; therefore they are on the same vertical line. To find their distance, subtract the $y$-coordinates $-2 - 4 = -6$. Because a distance cannot be negative, we say that the distance is 6.

**Example 2:** Find the distance between points $(3, -3)$ and $(5, 1)$.

If the points are not on the same horizontal or vertical line, you must use the *distance* formula to find how far apart they are. This formula is provided for you on the FCAT Reference Sheet. For points $(x_1, y_1)$ and $(x_2, y_2)$, the distance $d$ between the two points is found using this formula:

$$d = \sqrt{(x_2 - x_1)^2 + (y_2 - y_1)^2}$$

$$d = \sqrt{(5 - 3)^2 + (1 - (-3))^2}$$

$$d = \sqrt{(2)^2 + (4)^2}$$

$$d = \sqrt{4 + 16} = \sqrt{20} \approx 4.5$$

The answer is irrational and is rounded to 4.5.

## SLOPE

The slope of a line tells both how steep the line is and what its direction is. A line with positive slope travels upward from left to right, while a line with negative slope travels downward from left to right.

**Example:** On the graph shown, line $\overleftrightarrow{AB}$ has positive slope and $\overleftrightarrow{CD}$ has negative slope.

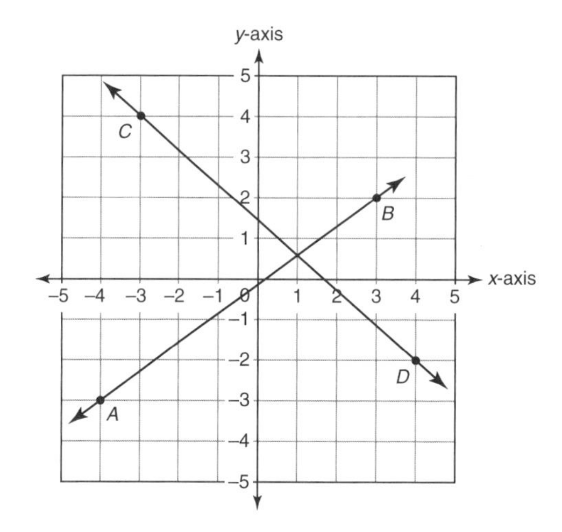


The steepness of a line is measured by finding $\frac{\text{vertical change}}{\text{horizontal change}}$. This is sometimes called $\frac{\text{rise}}{\text{run}}$ or "rise over run" to help you remember it.

**Example:** Find the slope of the line containing points (–2, 4) and (5, 9).

The vertical change of the line is the *difference* between the *y*-coordinates: 9 – 4 = 5. The horizontal change of the line is the *difference* between the *x*-coodinates: 5 – (–2) = 7. The slope is $\frac{5}{7}$.

The formula for slope is written as:

$$m = \frac{y_2 - y_1}{x_2 - x_1}.$$

The letter *m* is the symbol used for slope. The $y_2 - y_1$ portion of the formula represents the vertical change. The $x_2 - x_1$ portion represents the horizontal change. The higher the absolute value for slope, the steeper the line. A line with slope of $\frac{-2}{1}$ or –2 (absolute value: 2) is steeper than a line with slope of $\frac{1}{2}$.

**Example:** Find the slope of the line containing points *E* and *F* on the graph.

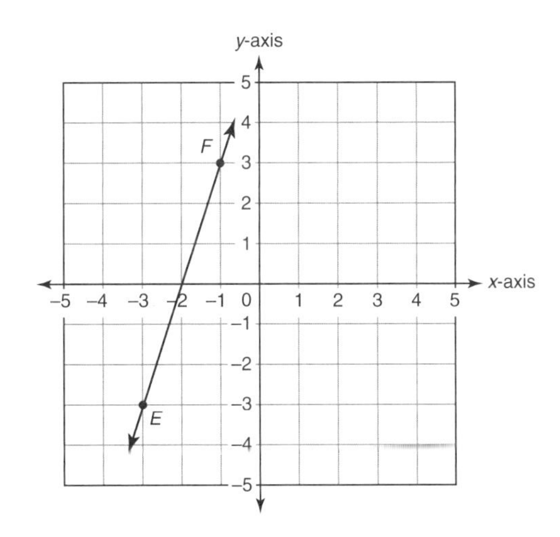


The points shown are *E*(–3, –3) and *F*(–1, 3). Use the slope formula:

$$m = \frac{y_2 - y_1}{x_2 - x_1} = \frac{3-(-3)}{-1-(-3)} = \frac{3+3}{-1+3} = \frac{6}{2} = 3.$$

## Horizontal Lines

*Horizontal lines* have zero slope because there is no rise: $\frac{\text{rise}}{\text{run}} = \frac{0}{\text{run}} = 0$.

**Example:** Points $K(-4, 3)$ and $L(3, 3)$ lie on a horizontal line, as shown on the graph. The slope is

$$m = \frac{y_2 - y_1}{x_2 - x_1} = \frac{3-3}{3-(-4)} = \frac{0}{7} = 0.$$

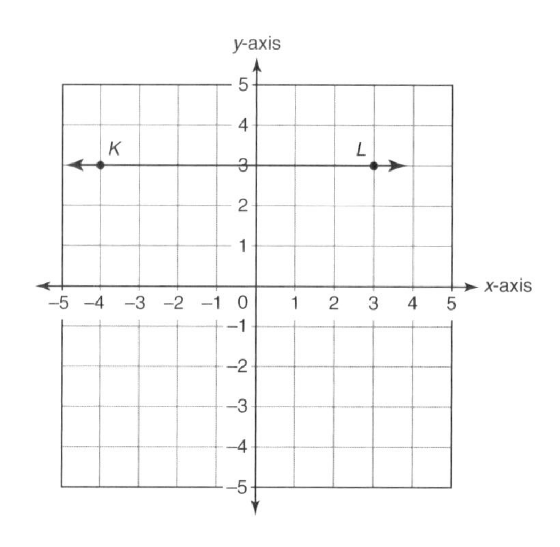


## Vertical Lines

*Vertical lines* have undefined slope because $\frac{\text{rise}}{\text{run}} = \frac{\text{rise}}{0}$ = undefined.

**Example:** Points $M(1, 4)$ and $N(1, -3)$ lie on a vertical line, as shown on the graph. The slope is

$$m = \frac{y_2 - y_1}{x_2 - x_1} = \frac{-3-4}{1-1} = \frac{-7}{0} = \text{undefined}.$$

**Division by zero is always undefined.**

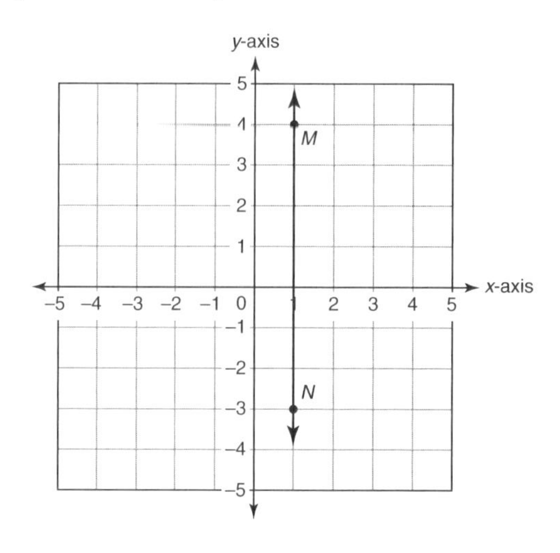

## Parallel Lines

*Parallel lines* have the same slope, as shown on the graph.

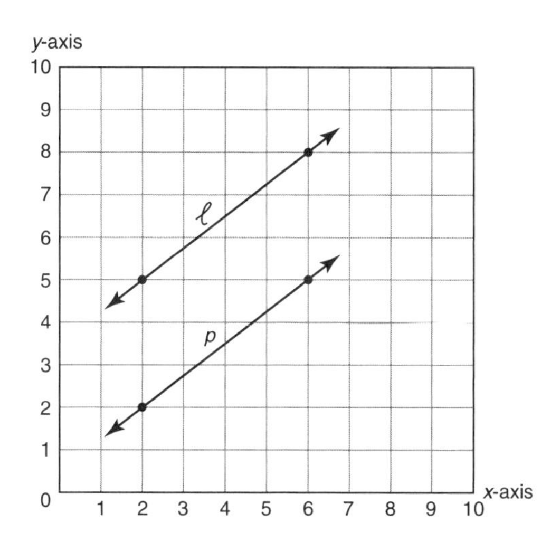


Line $\ell$ contains points (2, 5) and (6, 8). Its slope is

$$m = \frac{8-5}{6-2} = \frac{3}{4}.$$

Line $p$ contains points (2, 2) and (6, 5). Its slope is

$$m = \frac{5-2}{6-2} = \frac{3}{4}.$$

The slopes of the two lines are equal; therefore, the two lines are parallel ($\|$): $\ell \| p$.

## Perpendicular Lines

*Perpendicular lines* travel in opposite directions. Therefore, one line has negative slope and the other has positive slope. The slopes are opposite reciprocals of each other. For example, if the slope of one line is $-\frac{2}{3}$, the slope of its perpendicular is $+\frac{3}{2}$.

**Example:** On the graph shown, line $p$ contains points (–3, –1) and (1,1). Line $q$ contains points (–2, 2) and (0, –2).

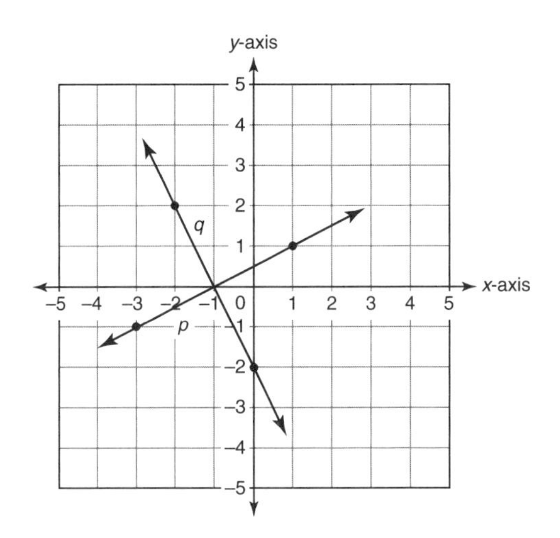


The slope of line $p$ is $m = \frac{1-(-1)}{1-(-3)} = \frac{1+1}{1+3} = \frac{2}{4} = \frac{1}{2}$.

The slope of line $q$ is $m = \frac{-2-2}{0-(-2)} = \frac{-4}{2} = -2$.

Since $p$ and $q$ are opposite reciprocals of each other, the two lines are perpendicular ($\perp$): $p \perp q$.

## INTERCEPTS

The point at which a line crosses either the $x$-axis or the $y$-axis is called an *intercept*. The coordinates of the $x$-intercept are $(x, 0)$, and the coordinates of the $y$-intercept are $(0, y)$.

**Example:** Find the $x$- and $y$-intercepts of the line shown.

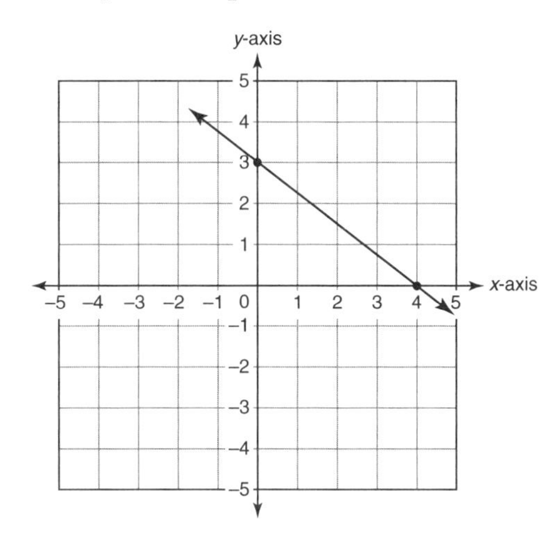


The $x$-intercept is $(4, 0)$. The $y$-intercept is $(0, 3)$.

## SLOPE-INTERCEPT EQUATIONS

The equation of a line in *slope-intercept* form is $y = mx + b$. The letter $m$ stands for the slope, and $b$ stands for the $y$-intercept. A line with an equation of $y = 2x - 3$ has a slope of $2\left(+\frac{2}{1}\right)$ and crosses the $y$-axis at $(0, -3)$, as shown.

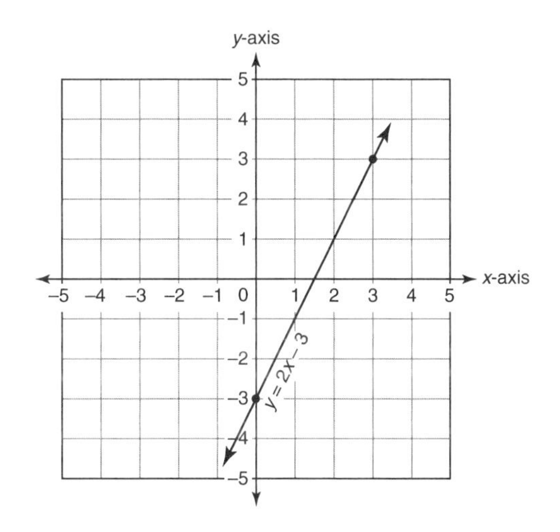

# SAMPLE QUESTIONS

## Coordinate Geometry—Distance

1. What are the coordinates of the other points needed to complete the vertices of a parallelogram?

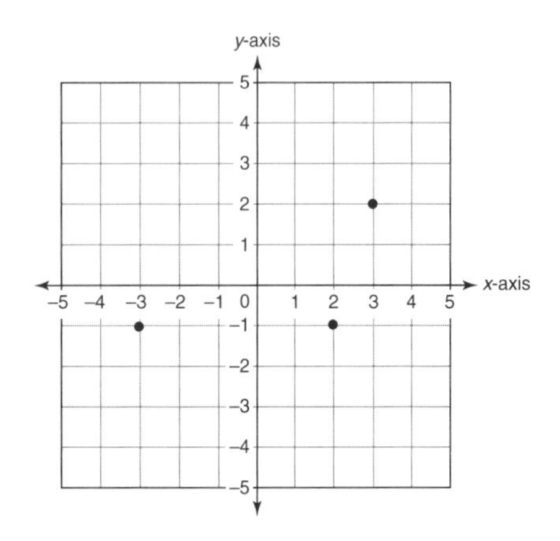


A. (2, –2)
B. (–2, 2)
C. (–1, 2)
D. (2, –1)

2. What are the coordinates of the intersection of State Street and Fannin Avenue?

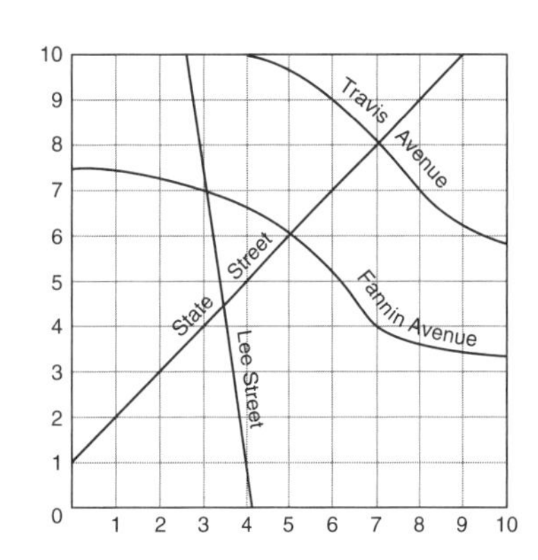


F. (6, 5)
G. (5, 1)
H. (6, 1)
I. (5, 6)

3. Find the distance between points *A* and *B*.

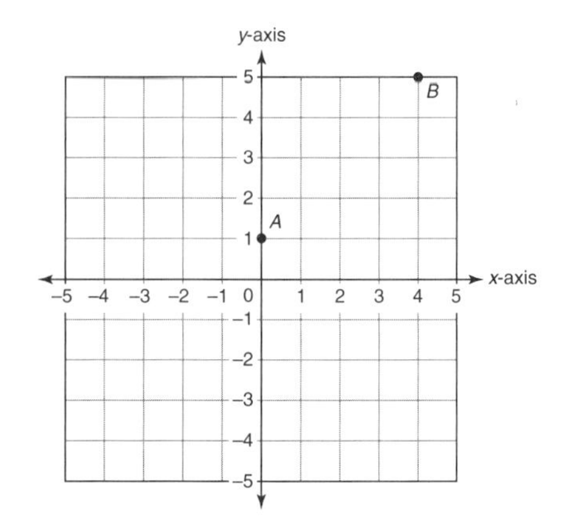


4. Find the midpoint between *A*(–7, 2) and *B*(–1, 4).

A. 6, –8
B. –4, 3
C. 2, 6
D. 1, 3

5. Find the slope of the line connecting points *A* and *B*.

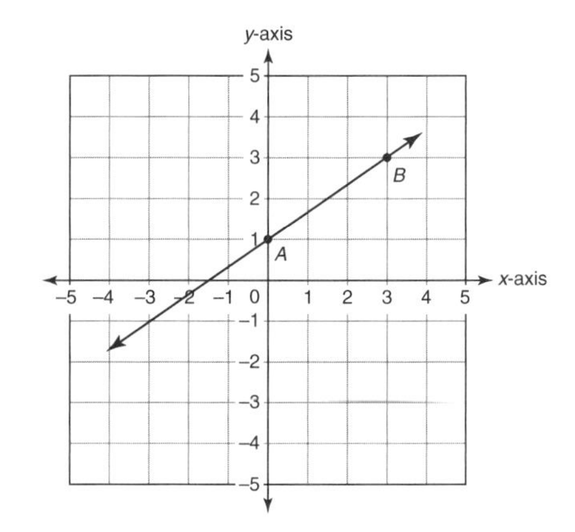

6. Find the $y$-intercept of line $AB$.

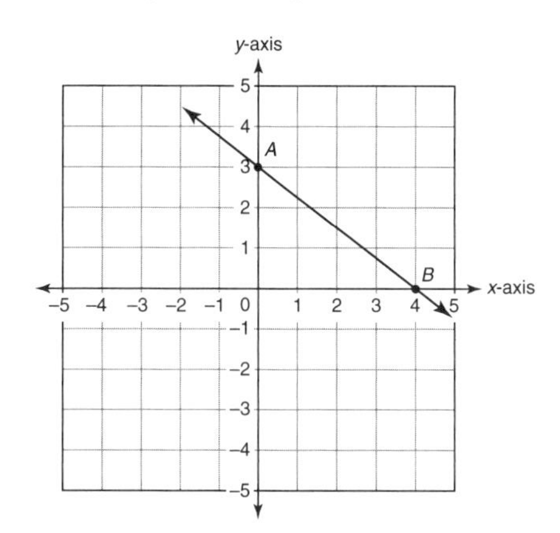


F. (3, 0) H. (0, 3)
G. (4, 0) I. (0, 4)

7. Find the $x$-intercept of line $CD$.

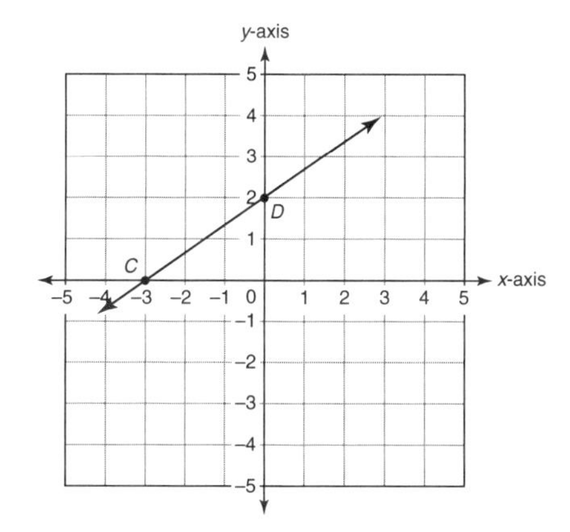


A. (2, 0) C. (–3, 0)
B. (0, 2) D. (0, –3)

8. Line $k$ has a slope of $\frac{8}{12}$. Which could be the slope of a line that is parallel to line $k$?

F. $-\frac{2}{3}$ H. $\frac{2}{3}$

G. $\frac{3}{2}$ I. $-\frac{3}{2}$

9. Nancy gives you directions from your house to her party. She tells you to go three blocks east and four blocks north. If you could go in a straight line, how many blocks would you travel from your house to Nancy's house?

10. What is the slope $m$ of the line parallel to a line containing points (–5, 1) and (4, 3)?

11. What is the slope $m$ of the line perpendicular to a line containing points (1, 0) and (0, 6)?

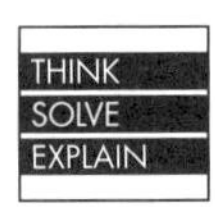


12. Line $CD$ is drawn on the graph.

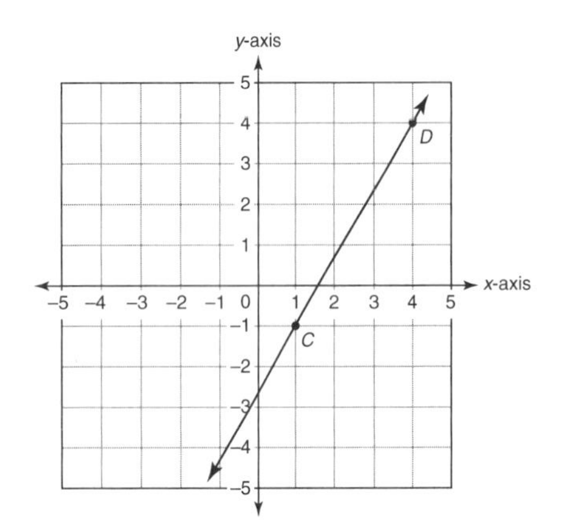


Part A. Graph the line passing through points (–3, –1) and (0, 4).

Part B. Is your line parallel to line CD? How do you know?

______________________________

______________________________

______________________________

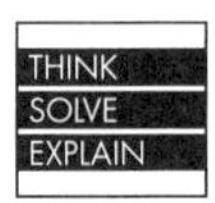


13. Line $RS$ is drawn on the graph.

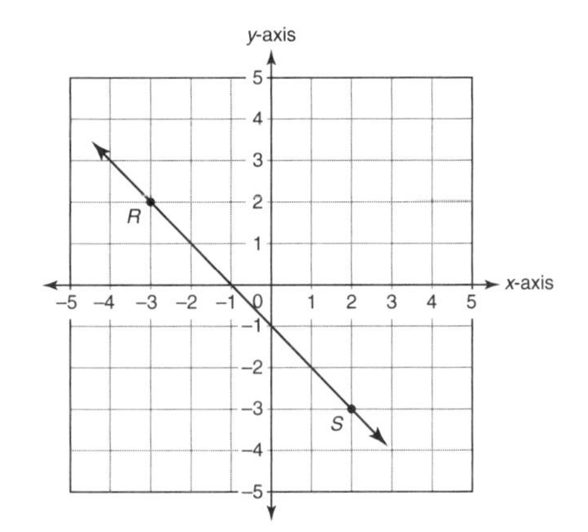


Part A. Graph the line passing through points (–2, –1) and (2, 4).

Part B. Is your line perpendicular to line $RS$? How do you know?

______________________________

______________________________

______________________________

# ANSWERS TO SAMPLE QUESTIONS

1. **B.** In a parallelogram, opposite sides are parallel. Therefore, the slopes of opposite sides are the same.

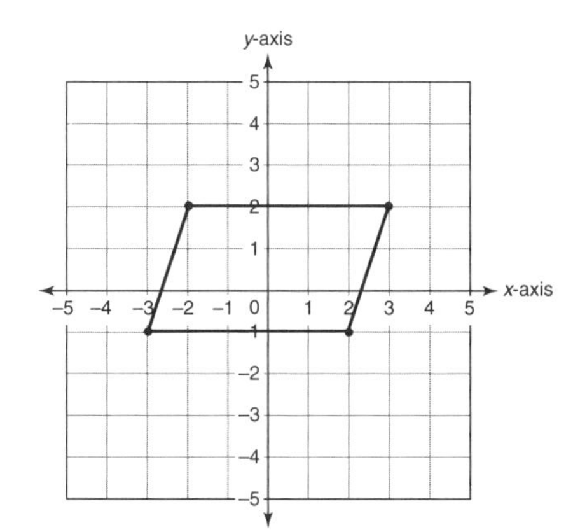


2. **I.** State Street and Fannin Avenue intersect where the $x$-coordinate is 5 and the $y$-coordinate is 6, that is, at point (5, 6).

3. **5.7.** The two points are not located on the same vertical or horizontal line. Use the distance formula:

$$d = \sqrt{(x_2 - x_1)^2 + (y_2 - y_1)^2}$$

$$= \sqrt{(4-0)^2 + (5-1)^2}$$

$$= \sqrt{4^2 + 4^2} = \sqrt{32} \approx 5.7.$$

4. **B.** Use the midpoint formula:

$$\frac{(x_1 + x_2)}{2}, \frac{(y_1 + y_2)}{2}$$

$$= \frac{[-7 + (-1)]}{2}, \frac{(2+4)}{2}$$

$$= \left(\frac{-8}{2}, \frac{6}{2}\right) = (-4, 3).$$

5. **$\frac{2}{3}$.** The slope is the vertical distance over the horizontal distance. $B$ is 2 spaces up from $A$ (rise) and 3 spaces to the right (run).

6. **H.** The $y$-intercept is the point at which the line crosses the $y$-axis, that is, point (0, 3).

7. **C.** The $x$-intercept is the point at which the line crosses the $x$-axis, that is, point (–3, 0).

8. **H.** Parallel lines have the same slope, and $\frac{8}{12}$ is equivalent to $\frac{2}{3}$.

9. **5.** The Pythagorean theorem and the distance formula are closely related. You can use the Pythagorean theorem to solve this problem.

$$a^2 + b^2 = c^2$$
$$3^2 + 4^2 = c^2$$
$$9 + 16 = c^2$$
$$25 = c^2$$
$$\sqrt{25} = 5 = c$$

10. **$\frac{2}{9}$.** Use the slope formula and substitute the given values:

$$m = \frac{y_2 - y_1}{x_2 - x_1}$$

$$m = \frac{3-1}{4+5} = \frac{2}{9}$$

11. $\frac{1}{6}$. Use the slope formula and substitute the given values:

$$m = \frac{y_2 - y_1}{x_2 - x_1}$$

$$m = \frac{6-0}{0-1} = \frac{6}{-1}$$

Perpendicular slopes are opposite reciprocals, so a line perpendicular to a line with a slope of –6 has a slope of $\frac{1}{6}$.

12. Part A.

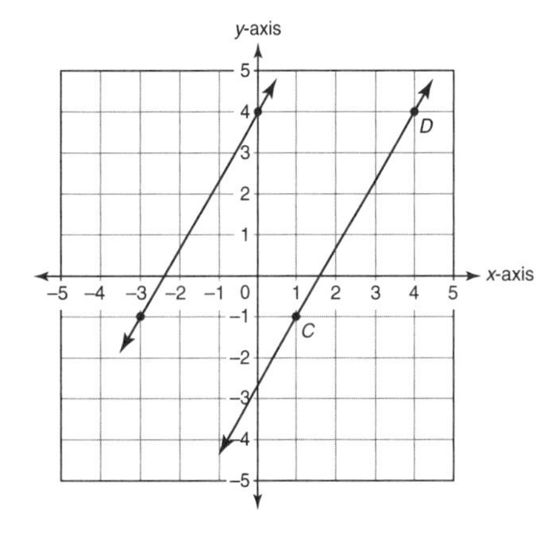


Part B. **Yes,** the lines are parallel. The slopes of the lines are the same: $\frac{5}{3}$.

13. Part A.

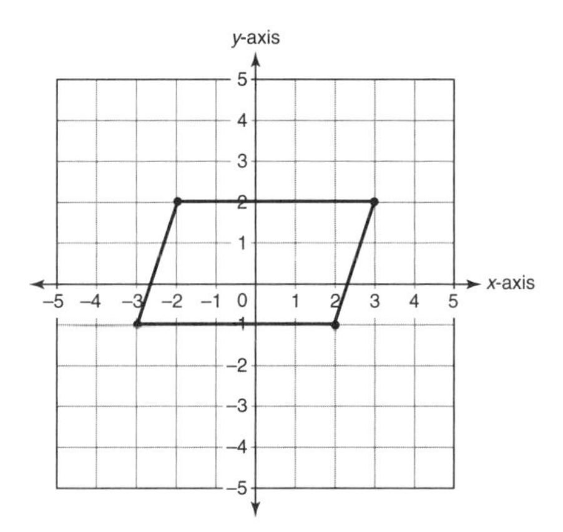


Part B. **No,** the lines are not perpendicular. The slope of line *RS* is –1, while the slope of the line containing points (–2, –1) and (2, 4) is $\frac{5}{4}$. These two numbers are not opposite reciprocals of each other.

# Lesson 16

# Line Symmetry and Transformations

## LINE SYMMETRY

Which of these figures shows a correctly drawn line of symmetry?

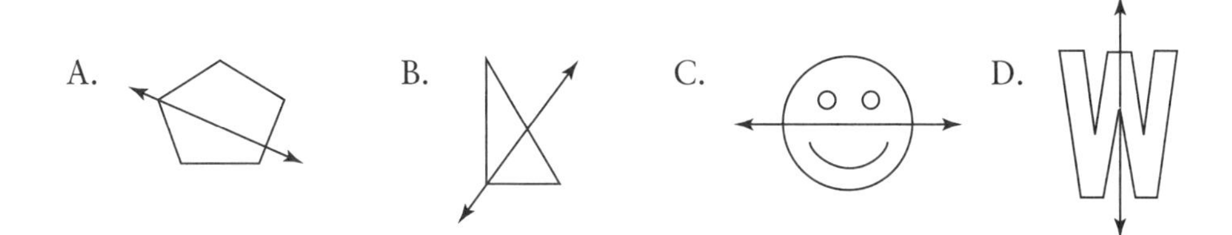


The only figure in which the two sides are mirror images is D. If you fold along a line of symmetry, the two sides will match or "map" onto each other.

A *line of symmetry* is a line that can be drawn through a plane figure so that the part of the figure on one side of the line is the congruent mirror image of the part on the other side of the line.

How many lines of symmetry are in a square?

As shown in the drawing, there are six lines of symmetry in a square.

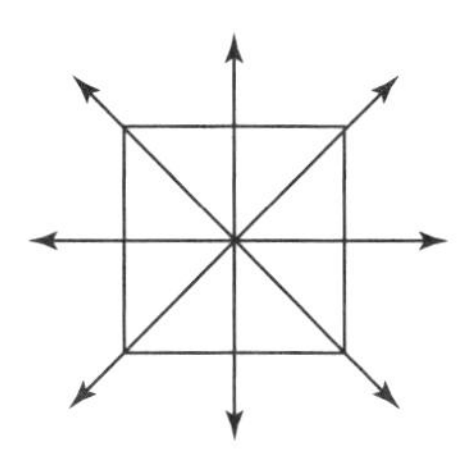

## TRANSFORMATIONS

To transform a figure, it must be moved. There are three basic *transformations*: (1) translation (slide), (2) reflection (mirror image or flip), and (3) rotation (turn). Before a figure is transformed, it is called the *preimage*; after it has been transformed, it is called the *image*. The image and preimage are congruent. In other words, transformations do not affect the size or shape of a figure.

## Translation (slide)

A *translation* slides all points of a figure the same distance. The figure can move up, down, right, left, or diagonally. When a figure is translated, its orientation is not changed. The preimage is congruent to the image.

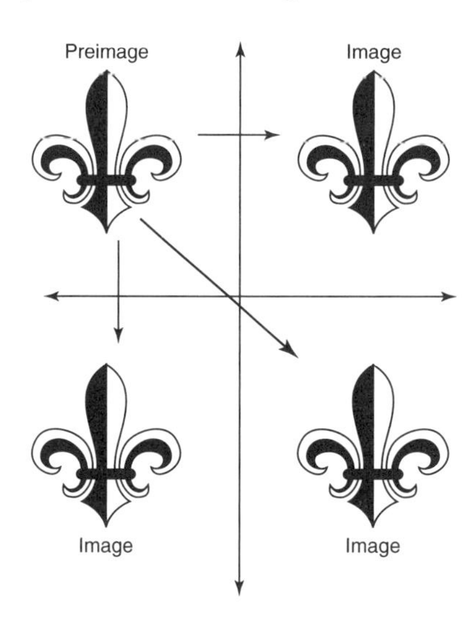


## Reflection (Flip or Mirror Image)

A *reflection* is a mirror image of a figure. As shown in the drawing, the figure is flipped over a line, forming a mirror image of the original figure. If the reflection is done correctly, you can fold along the line and both figures will match exactly. Each point on the image is exactly the same distance from the line of reflection as each point on the preimage.

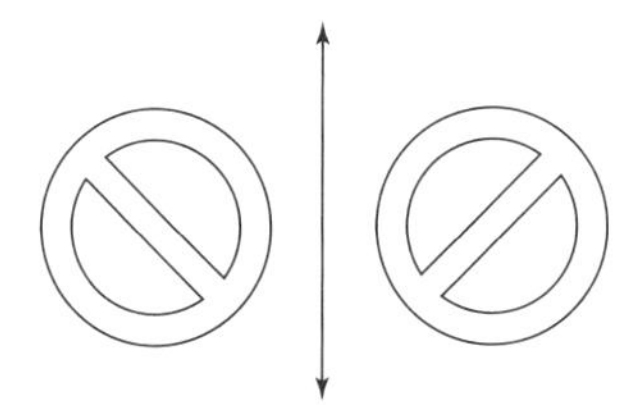

In a reflection, the image is facing the direction opposite to that of the preimage, as shown.

One way to tell whether you have reflected a figure properly is to draw a line from each point on the preimage to a corresponding point on the image. If your line forms a right angle with the reflecting line, you have done the reflection correctly.

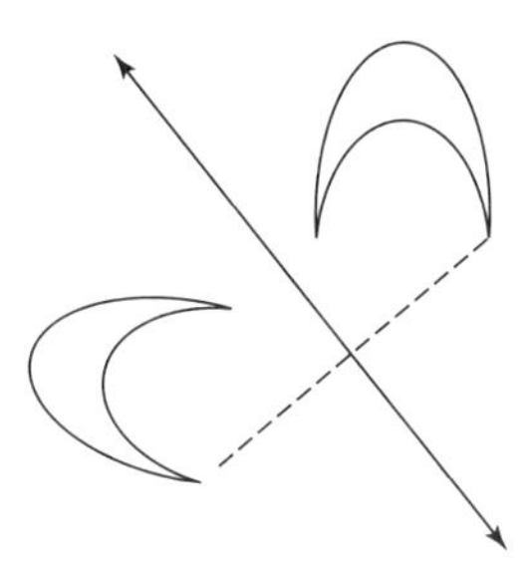

In the drawing, the darker figure, or preimage, is triangle *ABC*. In this case, the triangle is straddling the line. Notice that, when it is reflected over the line, its image is facing the opposite way and is also straddling the line. Each vertex of the image, or triangle *A′B′C′* (pronounced *A*-prime, *B*-prime, *C*-prime), is exactly the same distance from the reflecting line as each vertex of the darker preimage.

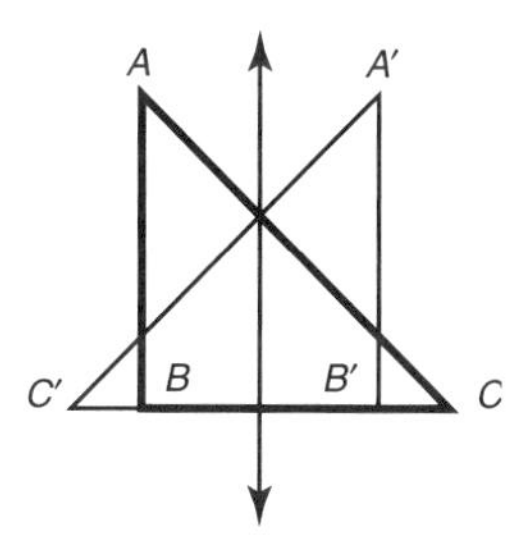


## Rotation (Turn)

A *rotation* is sometimes called a *turn*. In this transformation, the figure is turned about a point. You can imitate this transformation by cutting out a triangle and setting it down on a table. Put your pencil point on the triangle, and spin the triangle. You are performing a rotation. The tip of your pencil is the point about which the triangle is rotating.

In the diagram, trapezoid *ABCD* has been rotated 90 degrees clockwise. The center of rotation, or the point about which it was rotated, is the origin, or point *D*. The image of the trapezoid is trapezoid *A′B′C′D′*.

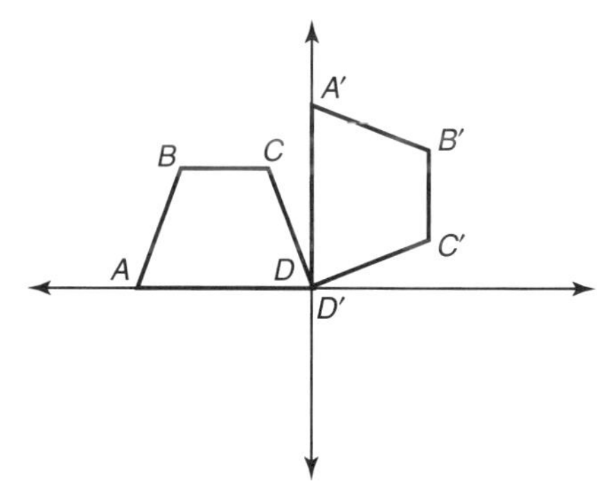

In the next drawing, Trapezoid *ABCD* has been rotated 180 degrees clockwise. Again, the center of rotation, or the point about which it was rotated, is the origin, or point *D*. The image of the trapezoid is trapezoid *A′B′C′D′*.

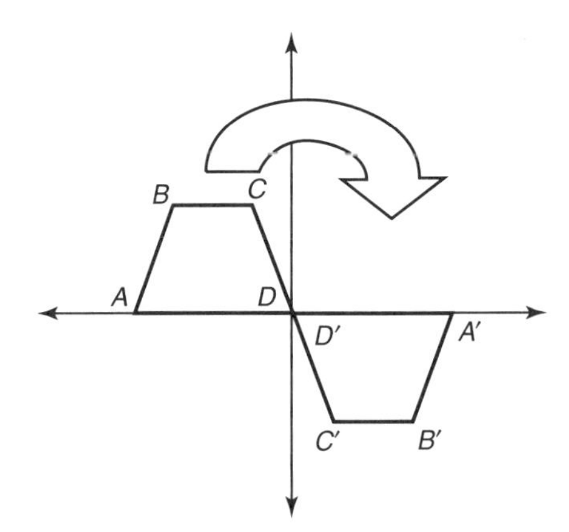


If a slide and a flip, instead of a rotation, had been performed, the effect would have been the same as that of a 180-degree rotation.

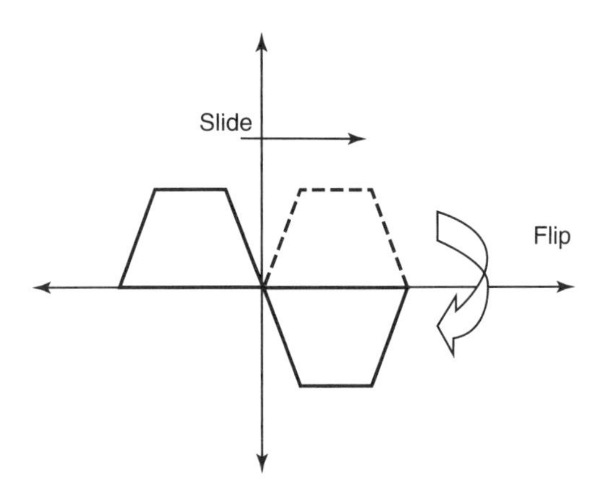

# SAMPLE QUESTIONS

## Line Symmetry and Transformations

1. What transformation will put this figure into proper form?

2. Which two transformations can change the preimage into the image?

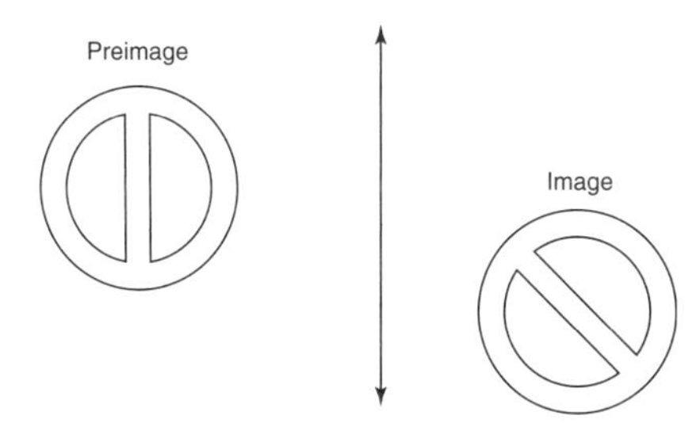


A. reflection and translation
B. rotation and reflection
C. rotation and translation
D. translation and symmetry

3. If the flag shown is reflected over the $y$-axis, at what point will A′ be located?

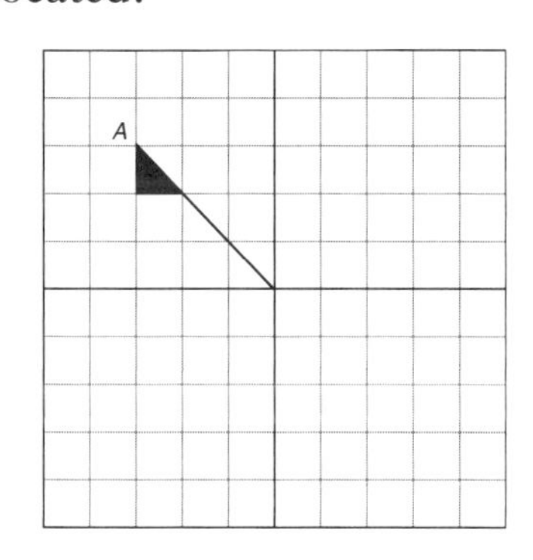


F. (3, –3)
G. (–3, –3)
H. (–3, 3)
I. (3, 3)

4. Reflect the figure over the $x$-axis. Then label the vertices of the image.

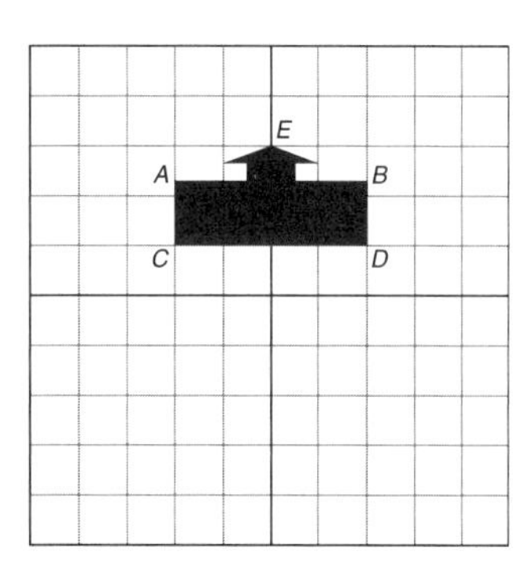


5. Draw a translation four spaces right and three spaces down of the trapezoid shown.

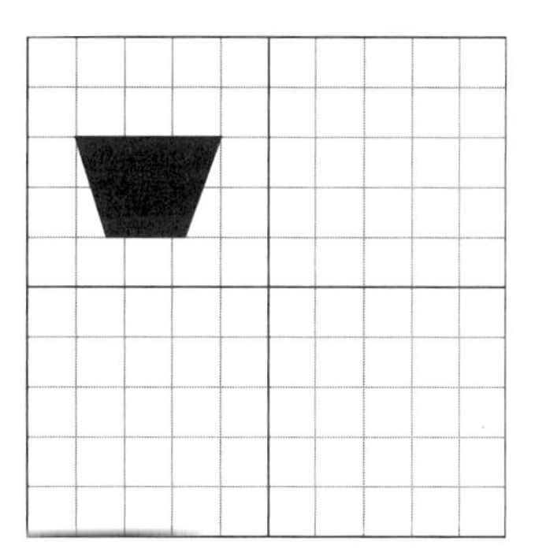

6. How many lines of symmetry are in a circle?

A. 2
B. 4
C. 8
D. an infinite number

## ANSWERS TO SAMPLE QUESTIONS

1. **Reflection** or **flip.** A mirror held to the right of the figure will reflect the word *Algebra*.

2. **C.** The preimage must be rotated and translated. Rotation and reflection will not work because the image and the preimage are not the same distance from the line.

3. **I.**

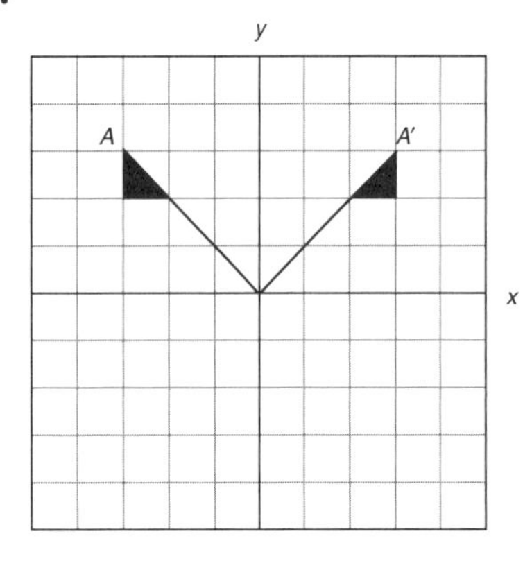


4.

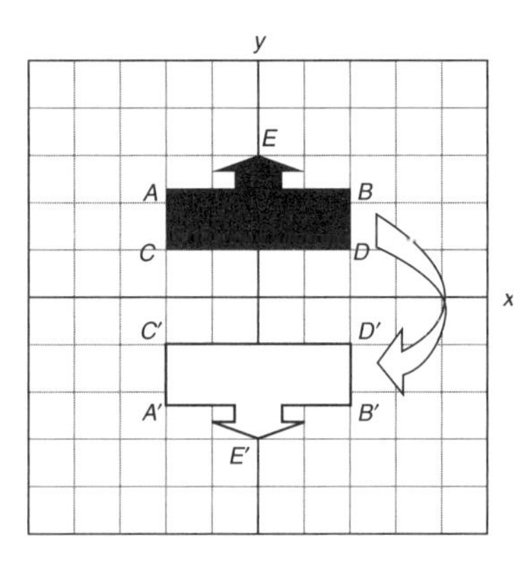


5.

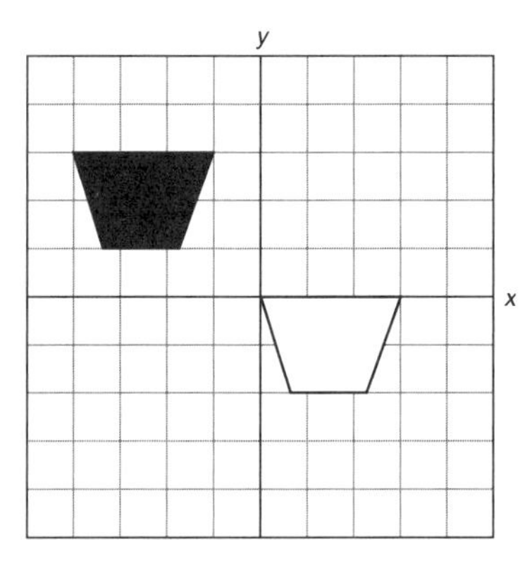


6. **D.**

# Lesson 17

# Patterns, Relationships, and Functions

What figure should be next in the set shown?

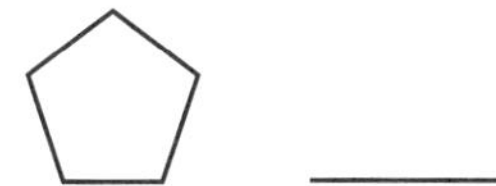

In this case, the pattern involves the number of sides in the figure: three, four, and five. We would expect the next figure to have six sides (a hexagon). There is also a secondary pattern; every other figure is shaded.

The hexagon would be shaded.

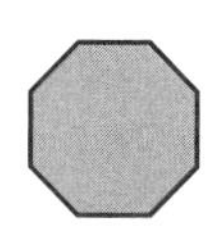

Patterns can be used to draw conclusions about mathematics or about things that happen in the real world.

## VISUAL PATTERNS

*Visual patterns* involve pictures that repeat in some predictable way.

**Example:**

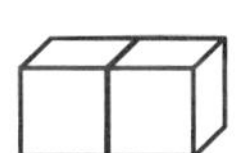

The triangular-shaped pattern of numbers shown below is called *Pascal's triangle*. Each row begins and ends with the number 1. Every other number is the sum of the two numbers directly above it. Can you fill in the last row by observing the pattern?

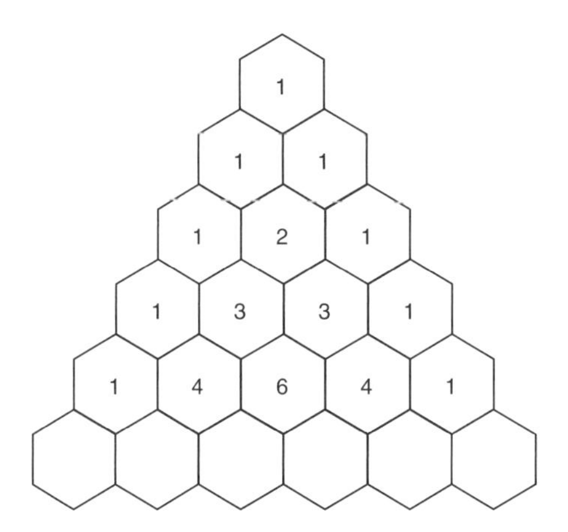


The last row should be 1, 5, 10, 10, 5, 1. Put 1 in each of the end blocks, and then fill in each of the other blocks by adding the two blocks directly above it. For example, 1 + 4 = 5 for the second block.

The next pattern is a famous one called the *Fibonacci sequence*. Below, the sequence is expressed visually using hearts. How many hearts should be next?

Because all the figures are the same, the pattern must be expressed in the number of hearts: 1, 1, 2, 3, 5, 8, ....

This is not an easy pattern to see at first, but careful inspection reveals that adding the first two terms gives the third term: 1 + 1 = 2.

Now, add the second term to the third term: 1 + 2 = 3, which is the fourth term.

Add the third term to the fourth term: 2 + 3 = 5.

Add the fourth term to the fifth term: 3 + 5 = 8.

By continuing in this way, you find that the seventh term is 5 + 8 = 13.

## Arithmetic Sequence

An *arithmetic sequence* is a pattern of three or more numbers that share a common difference. In this type of sequence each term is formed by adding a constant to the preceding term. This constant is the *common difference*. Remember that the word *difference* means to subtract.

**Examples:** 4, 5, 6, 7, ... has a common difference of 1.
5, 10, 15, ... has a common difference of 5.
10, 20, 30, ... has a common difference of 10.

**Example:** 10, 12, 14, 16 is an arithmetic sequence.

The common difference is 2 because 12 – 10 = 2; 14 – 12 = 2; and 16 – 14 = 2.

The fourth term is 16. You get this term by adding the common difference, 2, to the first term (10) three times:

$$10 + (2 \times 3) = 16.$$

If you want the 99th term, add 2 to the first term 98 times:

$$10 + (2 \times 98) = 10 + 196 = 206.$$

For the 200th term, add 2 to the first term 199 times:

$$10 + (2 \times 199) = 10 + 398 = 408.$$

## Geometric Sequence

A *geometric sequence* is a pattern of three or more numbers that share a common ratio. A common ratio can be found by dividing any term by the term immediately preceding it.

**Example:** The geometric sequence is 10, 20, 40, 80.

The common ratio is: $\frac{20}{10} = 2$; $\frac{40}{20} = 2$; $\frac{80}{40} = 2$. Each term is found by multiplying the preceding term by 2.

*Test Tip: Always check at least three ratios to be certain the pattern is a geometric sequence.*

**Example 1:** What is the eighth term in the geometric sequence 4, 12, 36, 108?

First find the common ratio: $\frac{12}{4} = 3; \frac{36}{12} = 3; \frac{108}{36} = 3.$

The common ratio is 3.

To find the eighth term in the sequence, use your calculator.

*Calculator Use for Geometric Sequences*

| Keystroke | Display | Term |
|---|---|---|
| [3] [×] [4] [=] | 12 | 2nd |
| [=] | 36 | 3rd |
| [=] | 108 | 4th |
| [=] | 324 | 5th |
| [=] | 972 | 6th |
| [=] | 2916 | 7th |
| [=] | 8748 | 8th |

The eighth term in the geometric sequence 4, 12, 36, 108 is 8748.

**Example 2:** What is the sixth term in the geometric sequence 64, 32, 16, 8?

First find the common ratio: $\frac{32}{64} = 0.5; \frac{16}{32} = 0.5; \frac{8}{16} = 0.5.$

The common ratio is 0.5.

*Calculator Use for Geometric Sequences*

| Keystroke | Display | Term |
|---|---|---|
| [0.5] [×] [64] [−] | 32 | 2nd |
| [=] | 16 | 3rd |
| [=] | 8 | 4th |
| [=] | 4 | 5th |
| [=] | 2 | 6th |

The sixth term in the geometric sequence 64, 32, 16, 8 is 2.

# RELATIONSHIPS

*Relationships* are sets of ordered pairs representing two sets of information. In class you may have compared your height and your shoe size, or states and their capitals. Relationships can be expressed in symbols, variables, tables, and graphs.

## Symbols and Variables

A relationship can be described by using a symbol or variable to stand for something else. For example, the fact that a certain apple tree has 102 apples on it can be expressed by using the letter $a$ to stand for "apples": $a = 102$. This is also called an *algebraic expression.*

## Tables

To describe the relationship between the types of trees in an orchard and the numbers of fruit on the various trees, you might use a table.

| Type of Tree | Number of Fruit |
|---|---|
| Apple | 102 |
| Plum | 203 |
| Cherry | 109 |
| Peach | 80 |
| Pear | 174 |

Write the ordered pairs for this relationship:

(Apple, 102); (Plum, 203); (Cherry, 109); (Peach, 80); (Pear, 174).

The first item in each ordered pair is the domain of the relationship. The second item (a number in this case) in the ordered pair is the range.

## Graphs

The relationship shown for the orchard can be easily graphed. Because you are comparing the numbers of fruit on the various trees, a bar graph is the best way to show the relationship. The first item (type of tree) in each ordered pair is graphed on the horizontal axis, and the second item (number of fruit) is graphed on the vertical axis.

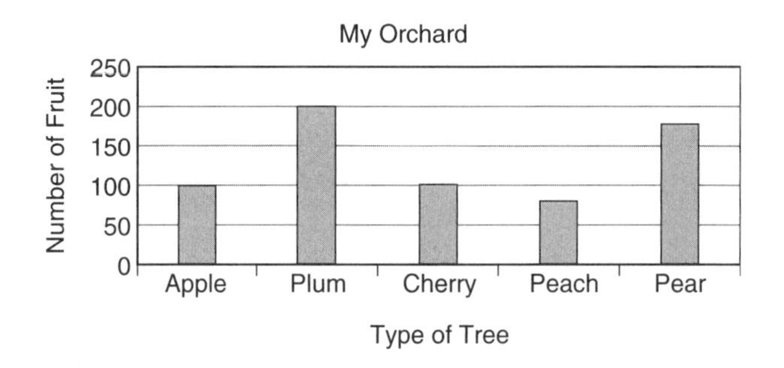


# FUNCTIONS

A *function* is a special type of relationship. Each element in the domain is paired with exactly one element in the range.

**Example:** Kara, Jason, and John have Mrs. Anderson as their science teacher, while Sheena, Beth, and Zach have Mr. Jones as their science teacher.

The relationship shown is a function because each element in the domain (student) has only one element in the range (teacher). Each student has only one teacher.

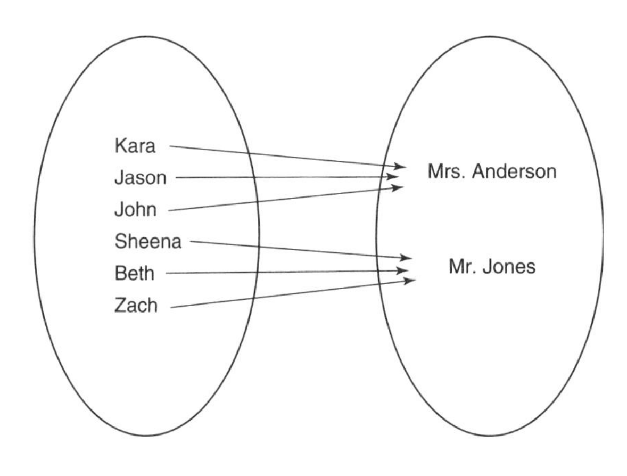


Domain (students)    Range (teachers)

If the roles are reversed, the relationship is no longer a function because each element in the domain (teacher) has more than one element in the range (student).

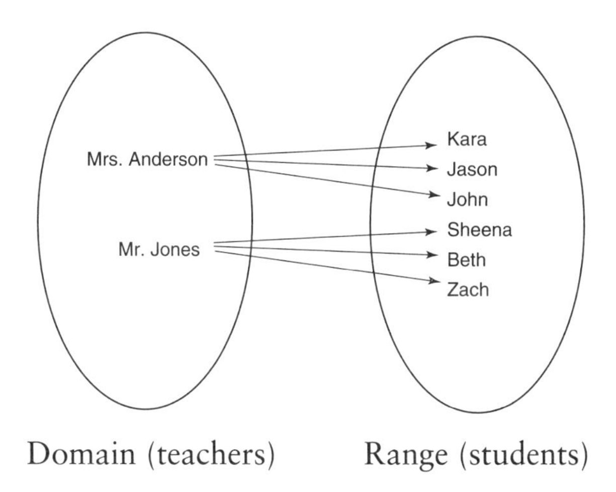


Domain (teachers) Range (students)

**Example:** The table shows the height of a tree each year for 5 years.

| Age of Tree | Height of Tree |
|---|---|
| 1 year | 3 feet |
| 2 years | 6 feet |
| 3 years | 12 feet |
| 4 years | 16 feet |
| 5 years | 22 feet |

1. Write the ordered pairs of this relationship (age, height).
2. Write the domain.
3. Write the range.
4. Is this relationship a function?

1. The ordered pairs for the relationship are (1, 3); (2, 6); (3, 12); (4, 16); and (5, 22).
2. The domain is {1, 2, 3, 4, 5}.
3. The range is {3, 6, 12, 16, 22}.
4. The relationship is a function.

# SAMPLE QUESTIONS

## Patterns, Relationships, and Functions

1. What is the next number in the sequence –10, –8, –6, –4?

2. If a row of balls is added at the bottom in the pattern shown, how many balls will be in that row?

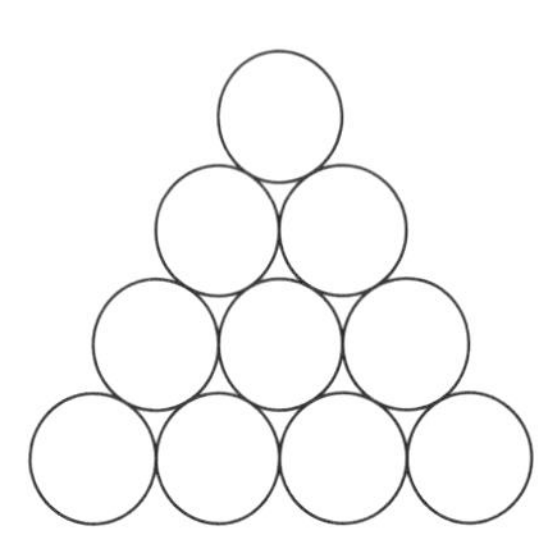

3. Find the fifth number in the sequence 1, 4, 9, 16, ____.

4. Fill in the numbers in the bottom row. Use your answer to fill in the blank at the end of the lowest arrow.

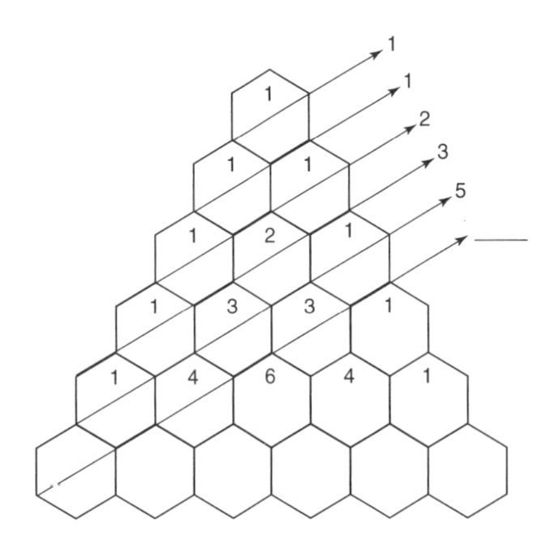


5. Which letter has more than one line of symmetry?

   A. **A**
   B. **E**
   C. **X**
   D. **Q**

6. Find the next number in the sequence 6, 12, 24, 48, ____.

7. Which of the following sets of ordered pairs is NOT a function?

   F. (1, 1); (2, 2); (3, 3); (4, 4)
   G. (0, 1); (0, 2); (1, 3); (1, 4)
   H. (1, 0); (2, 0); (3, 0); (4, 0)
   I. (–2, 5); (–4, –4); (–6, 10); (–8, 12)

8. The table shown represents a function. Identify the missing value.

| $x$ | 2 | 4 | 6 | 8 |
|---|---|---|---|---|
| $y$ | 5 | 7 | 9 | |

9. Some graph relationships are functions. One quick way to tell whether a graph is a function is by using the straight-line test. If a vertical line drawn through the graph does not touch the graph in more than one place, the relationship is a function. Use the straight-line test to identify the graph that is NOT a function.

A. 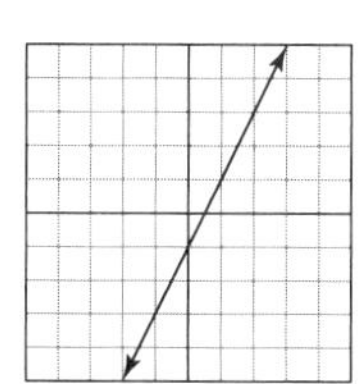

C. 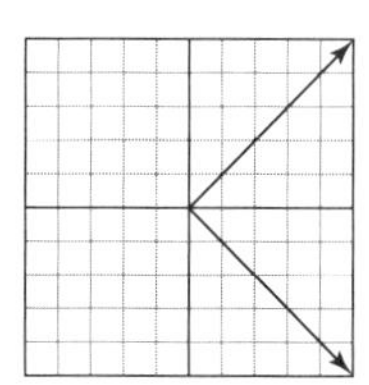

B. 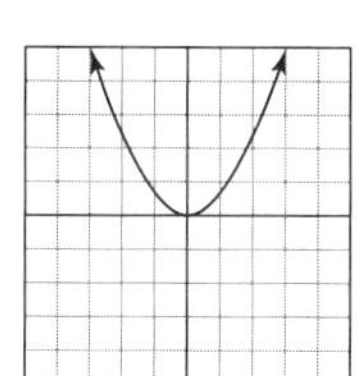

D. 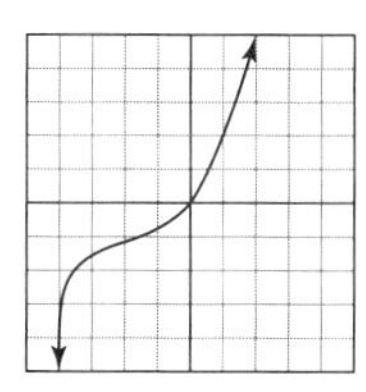

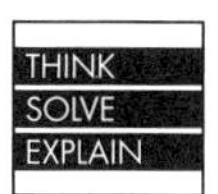


10. Part A. On the grid provided, graph this relationship: (–4, –2); (–3, 3); (0, 4); (1, 3); (1, –2); (2, 0).

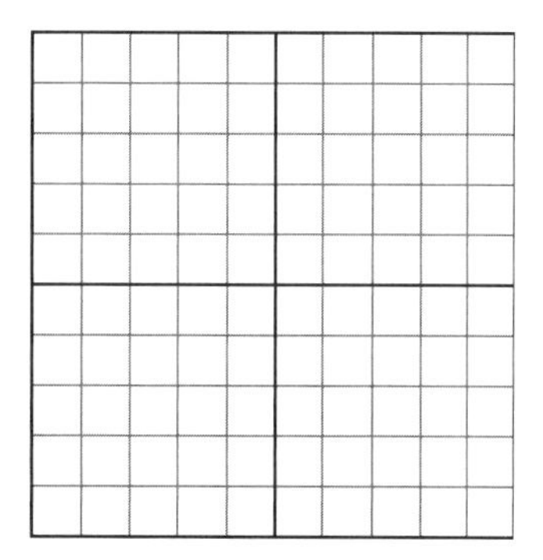

Part B. Is the relationship a function? Explain.

______________________________

______________________________

______________________________

# ANSWERS TO SAMPLE QUESTIONS

1. **–2.** This is an arithmetic sequence that is increasing by 2 each time.

2. **5.** The first row has 1 ball, the second row has 2 balls, etc.

3. **25.** Each number is a perfect square: $1 \times 1 = 1$; $2 \times 2 = 4$; $3 \times 3 = 9$; $4 \times 4 = 16$; $5 \times 5 = 25$.

4. **8.** The bottom row in the figure should be 1, 5, 10, 10, 5, 1. The numbers at the ends of the arrows are found by adding the numbers inside the hexagons through which the arrow passes. Therefore, $1 + 4 + 3 = 8$. This is another example of the Fibonacci sequence.

5. **C.** The letter X has two lines of symmetry. Each of the other letters has only one line of symmetry.

6. **96.** This is a geometric sequence in which each number is found by multiplying the preceding number by 2.

7. **B.** Each number in the domain {0, 1} can have only one number in the range {1, 2, 3, 4}. In B, 0 has two numbers in the range {1, 2} and 1 has two numbers in the range {3, 4}.

8. **11.** The pattern for the *y*-values is arithmetic and increasing by 2. Each *y*-value is 3 more than its corresponding *x*-value.

9. **C.** See the straight lines drawn through the graphs. The line touches graph C in two places but touches each of the other graphs in only one place.

10. Part A.

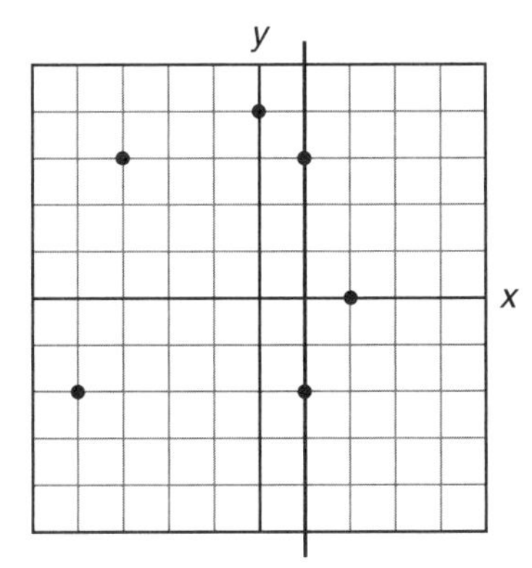


Part B. **No.** The relationship is not a function. For the domain $x = 1$, there are two ranges, $y = 3$ and $y = -2$. Also, a straight line drawn vertically can pass through two points, as shown in the drawing above.

# Lesson 18

# Variables, Expressions, Equations, and Inequalities

## VARIABLES AND EXPRESSIONS

A sales clerk's weekly salary is calculated by figuring the sum of $6.50 per hour and $50. Write an algebraic expression for the clerk's weekly salary.

The algebraic expression would look like this: $6.50h + 50$.

*Algebraic* expressions are used to translate English statements into equations. When a number is not known a letter called a *variable* is substituted for it. The variable $h$ has been substituted for the hourly wage. $6.50 should be multiplied by $h$ to find the total earned before adding $50.

### Writing Expressions

Some additional examples of English expressions changed to algebraic expressions are shown in the table below.

| English Expression | Algebraic Expression |
|---|---|
| A number increased by 2 | $n + 2$ |
| 3 more than a number | $n + 3$ |
| A number decreased by 10 | $n - 10$ |
| A number less 5 | $x - 5$ |
| Subtract 1 from a number | $a - 1$ |
| 7 less than a number | $a - 7$ |
| Twice a number | $2x$ |
| The product of a number and 9 | $9a$ |
| The sum of twice a number and 3 | $2n + 3$ |
| A number squared | $a^2$ |
| The area of a square with side length $n$ | $n^2$ |
| A number cubed | $a^3$ |
| The volume of a cube with side length $x$ | $x^3$ |
| Three times the sum of $x$ and 2 | $3(x + 2)$ |
| The sum of the squares of $a$ and $b$ | $a^2 + b^2$ |

**Example 1:** Write an expression for 5 less than three times a number.

Break it down: 5 less than means subtract 5.
Three times a number means multiply 3 by a variable (use $x$)

Put it together: $3x - 5$.

**Example 2:** Write an expression for the sum of twice a number and 31.

Break it down: Sum means add: "Twice a number" + 31.
Twice a number means multiply 2 by a variable (use $a$).

Put it together: $2a + 31$.

### Evaluating Expressions

Expressions are *evaluated* when a value is substituted for the variable.

**Example 1:** Write a variable expression for the phrase "the difference of 25 and $a$." Then evaluate the expression for $a = 20$.

The variable expression for the difference of 25 and $a$ is $25 - a$. If 20 is substituted for $a$, the expression equals $25 - 20$. Once the arithmetic is complete, the expression has been evaluated. The answer is 5.

**Example 2:** Write a variable expression for the phrase "the sum of $a$ and twice $b$." Evaluate the expression for $a = 4$ and $b = 5$.

The variable expression is $a + 2b$.
Substitute the given values: $4 + 2 \times 5 = 4 + 10 = 14$.

## EQUATIONS

You can use algebraic expressions to help you to write an equation. An *equation* is a mathematical sentence that combines two expressions by using an equal sign. When the equation is true, both sides are equivalent.

**Example:** Les mows lawns every week to earn extra money. He charges $15 per lawn. Write an equation for the total amount he earns in a week.

Connect: The total Les earns is 15 times the number of lawns he mows.
Translate: Let $T$ = total he earns.
Let $n$ = number of lawns he mows.

Write the equation: $T = 15n$.

Now suppose Les starts charging $20 per lawn. Write an equation to find the total amount he earns in a week.

The new equation is: $T = 20n$

## Writing Equations from Tables

What equation can you write to describe the relationship between the variables in the table?

| $x$ | $y$ |
|---|---|
| 5 | 10 |
| 10 | 20 |
| 20 | 40 |
| 30 | 60 |

First, look at each column. Notice that each value in the the $y$-column is twice as large as the corresponding value in the $x$-column. Translated: $y = 2x$, or $2x = y$.

**Example:** Use the table shown to write an equation.

| $b$ | $c$ |
|---|---|
| 1 | 6 |
| 2 | 7 |
| 3 | 8 |
| 4 | 9 |

Notice that each value of $c$ is 5 more than the corresponding value of $b$. Translated: $c = 5 + b$, or $c = b + 5$.

## Solving One-Step Equations

Equations are not evaluated; they are solved. To *solve* an equation, you find one or more values that make the equation true. The simplest method is to work backward by identifying what has been done to the variable and "undoing" it.

**Example 1:** $n - 12 = -42$. In this equation, 12 is subtracted from the variable. Because the opposite of subtracting is adding, simply add 12 to both sides of the equation. Remember to perform the addition to both sides so that the two sides maintain their balance.

| | |
|---|---|
| $n - 12 + 12 = -42 + 12$ | Add 12 to both sides. |
| $n = -30$ | Complete the arithmetic, and simplify. |

**Example 2:** The triangle shown is isosceles with leg $\overline{AB} \cong$ leg $\overline{BC}$. Find the value of $x$.

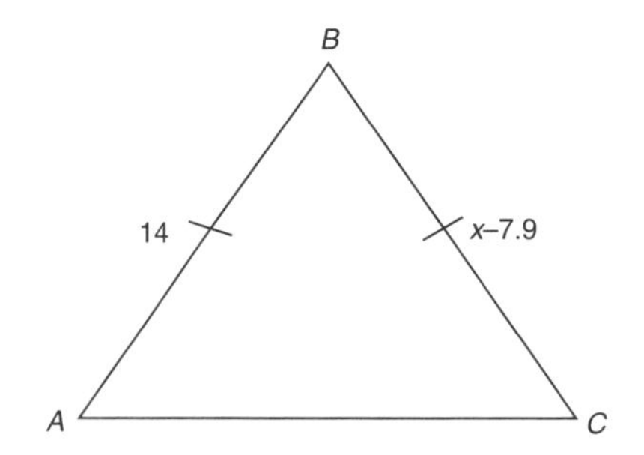


| | |
|---|---|
| $\overline{AB} \cong \overline{BC}$ | The legs of an isosceles triangle are congruent. |
| $14 = x - 7.9$ | Substitute given values. |
| $14 + 7.9 = x - 7.9 + 7.9$ | Add 7.9 to both sides. |
| $21.9 = x$ | Simplify. |

**Example 3:** Solve $3a = 54$ for $a$.

The variable is multiplied by 3. To solve, divide both sides by 3.

| | |
|---|---|
| $3a = 54$ | |
| $\frac{3a}{3} = \frac{54}{3}$ | Divide both sides by 3. |
| $a = 18$ | Simplify. |

**Example 4:** Solve $\frac{n}{-5} = 2$.

The variable is divided by –5. To solve, multiply both sides by –5.

$$\frac{n}{-5} = 2$$

$$\frac{n}{-5} \cdot -5 = 2 \cdot -5$$ Multiply both sides by –5.

$$n = -10$$ Simplify.

**Example 5:** Solve $\frac{3}{4}n = 12$. In an equation such as this, it helps to know that the variable is multiplied by 3 and divided by 4. You can "undo" this by using the fraction's reciprocal $\left(\frac{4}{3}\right)$ to multiply both sides of the equation.

$$\frac{3}{4}n = 12$$

$$\frac{3}{4} \cdot \frac{4}{3}n = 12 \cdot \frac{4}{3}$$ Multiply both sides by $\frac{4}{3}$.

$$n = 12 \cdot \frac{4}{3}$$ Simplify.

$$n = 12 \cdot 4 \div 3$$ Multiply by 4, and divide by 3.

$$n = 16$$

## Solving Two-Step Equations

Two-step equations require two operations to solve. The first operation requires you to use addition or subtraction. The second operation requires multiplication or division.

**Example 1:** Solve $20 = \frac{x}{3} + 4$.

$$20 = \frac{x}{3} + 4$$

$$20 - 4 = \frac{x}{3} + 4 - 4$$ Subtract 4 from both sides.

$$16 = \frac{x}{3}$$ Simplify.

$$48 = x$$ Multiply both sides by 3.

**Example 2:** In the triangle shown, $\angle A \approx \angle B$. Find the value of $x$.

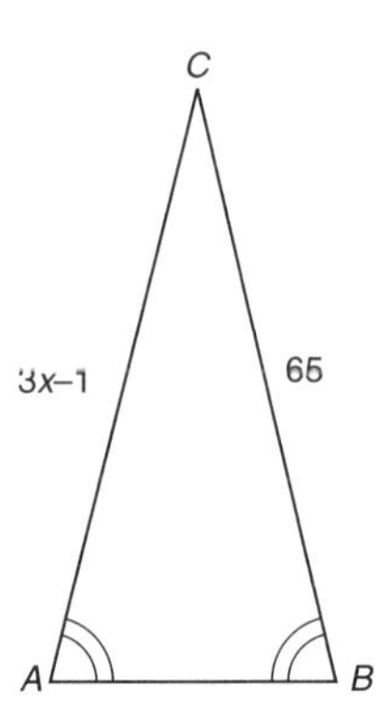


| | |
|---|---|
| $\angle A \approx \angle B$ | The measure of ∠A equals the measure of ∠B; therefore this triangle is isosceles. |
| $65 = 3x - 1$ | Substitute the given values. |
| $65 + 1 = 3x - 1 + 1$ | Add 1 to both sides. |
| $66 = 3x$ | Simplify. |
| $22 = x$ | Divide both sides by 3. |

## INEQUALITIES

Use algebraic notation to write this statement: Justin's salary is greater than $250 per week.

This statement is true if Justin earns any amount over $250 per week ($s > 250$). There are an infinite number of values that make this statement true (251, 251.50, 252, ...). For this reason, the statement is called an *inequality*. An inequality is a statement that compares the values of two expressions joined by an inequality symbol. The five inequality symbols are shown in the table.

| Symbol | Meaning | Algebraic | Solution to Expression |
|---|---|---|---|
| $<$ | Is less than | $n < 0$ | {... –3, –2, –1, –0.1} |
| $\leq$ | Is less than or equal to | $n \leq 5$ | {... 3, 4, 5} |
| $>$ | Is greater than | $n > -2$ | {–1, 0, 1, 2, ...} |
| $\geq$ | Is greater than or equal to | $n \geq 3$ | {3, 4, 5, ...} |
| $\neq$ | Is not equal to | $4 \neq -4$ | |

## Graphing Inequalities

A simple inequality with one variable is graphed on a number line, using an open or closed circle as indicated in the table and an arrow pointing in the direction of the solution.

| Inequality | Circle |
|---|---|
| $<$ | Open |
| $\leq$ | Closed |
| $>$ | Open |
| $\geq$ | Closed |

**Example 1:** Write and graph an inequality that says a number is greater than or equal to negative two.

Write: $n \geq -2$.

Graph:

**Example 2:** Write and graph an inequality that says Amy makes less than $7.50 per hour.

Write: $n < 7.50$

Graph:

**Example 3:** Write and graph an inequality that says tickets to a concert range in price from $35 to $60.

Write: $35 \leq t \leq 60$. Note that for a compound inequality the numbers are arranged from smaller to larger with the appropriate symbols.

Graph:

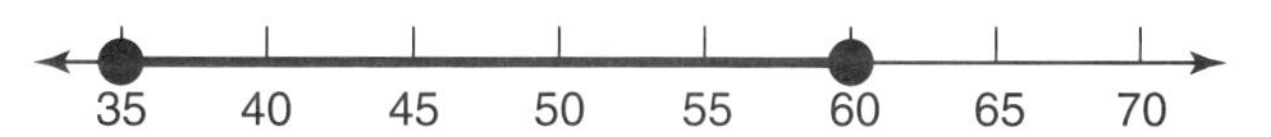

## Solving Inequalities

With one exception, inequalities are solved exactly like equations. The exception is: when multiplying or dividing by a negative number, the inequality is reversed.

**Example 1:** Solve, list three solutions, and graph the solution to $19 > n - 2$.

| | |
|---|---|
| $19 > n - 2$ | |
| $19 + 2 > n - 2 + 2$ | Add 2 to both sides. |
| $21 > n$ | Simplify. |
| $\{20, 19, 18\}$ | Since 21 is greater than $n$, only solutions less than 21 will work. |

Graph: –25 0 21 25

**Example 2:** Solve, list three solutions, and graph the solution to $2n + 2 < 10$.

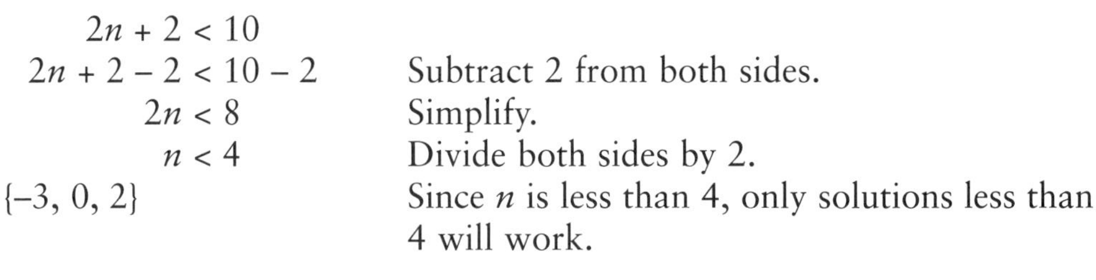

| | |
|---|---|
| $2n + 2 < 10$ | |
| $2n + 2 - 2 < 10 - 2$ | Subtract 2 from both sides. |
| $2n < 8$ | Simplify. |
| $n < 4$ | Divide both sides by 2. |
| $\{-3, 0, 2\}$ | Since $n$ is less than 4, only solutions less than 4 will work. |

Graph: –1 0 1 2 3 4 5 6

**Example 3:** Solve, list three solutions, and graph the solution to $-3x + 5 \le 20$.

| | |
|---|---|
| $-3x + 5 \le 20$ | |
| $-3x + 5 - 5 \le 20 - 5$ | Subtract 5 from both sides. |
| $-3x \le 15$ | Simplify. |
| $x \ge -5$ | Divide by –3, and REVERSE THE INEQUALITY SYMBOL! |
| $\{-5, -6, -7, -8, ...\}$ | Since $x$ is greater than or equal to –5, only solutions greater than or equal to –5 will work. |

Graph: –5 0 5 10

**Example 4:** Is the ordered pair (2, –3) a solution to the inequality $y < 2x - 1$?

| | |
|---|---|
| $y < 2x - 1$ | |
| $-3 < 2(2) - 1$ | Substitute the ordered pair (2, –3) for $x$ and $y$ in the inequality. |
| $-3 < 3$ | Simplify. |

Since it is true that –3 is less than 3, the ordered pair is one of the solutions to the inequality.

# SAMPLE QUESTIONS

## Variables, Expressions, Equations, and Inequalities

1. Which equation can be used to generate this table?

| $x$ | 2 | 4 | 6 | 8 |
|---|---|---|---|---|
| $y$ | –4 | 2 | 8 | 14 |

A. $y = x - 6$
B. $y = x + 6$
C. $y = 3x - 10$
D. $y = 3x + 4$

2. How can this graph be expressed as an inequality?

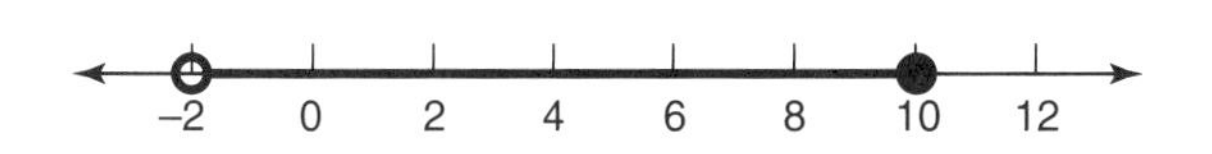


F. $-2 < x \leq 10$
G. $-2 \leq x \leq 10$
H. $-2 \leq x < 10$
I. $10 \leq x < -2$

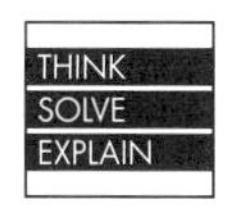


3. Forrest's age in 9 years will be less than 23.

Part A. Write an inequality in simplest form that models the above statement.

____________________________________

____________________________________

____________________________________

Part B. Solve the inequality from Part A, and graph below.

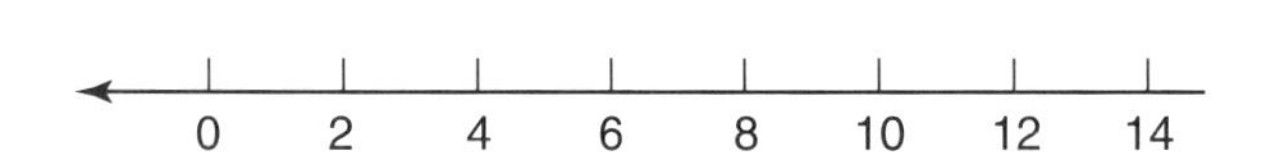


4. For the equation $y = 9x - 5$, what is the value of $y$ when $x = -3$?

5. The formula $d = rt$ is used to find distance $d$, rate $r$, or time $t$. If a car travels a distance of 600 miles in 8 hours, what is its average rate of speed in miles per hour?

6. Jay builds a ramp that rises 5 inches for every horizontal foot of ground it covers. If the ramp is 15 feet long, how high is the ramp?

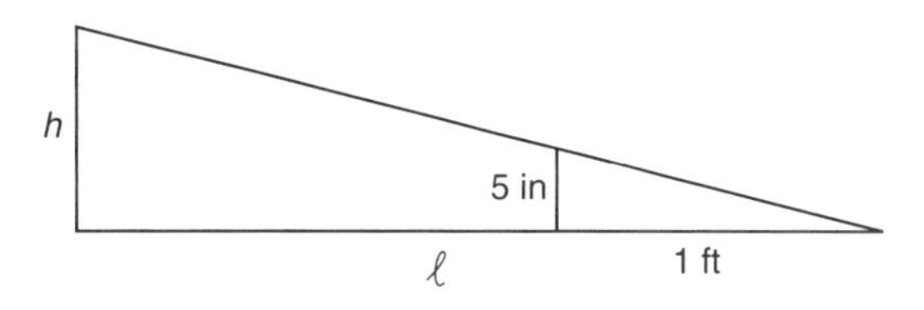


A. 6 ft
B. 6 ft, 3 in
C. 15 ft
D. 75 ft

7. For the sequence shown, which expression tells how to determine the number of seats from the number of tables $t$?

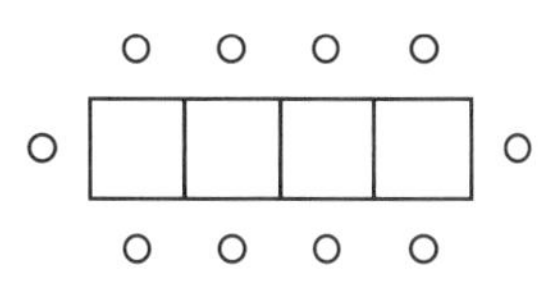

F. $4t$
G. $2t - 2$
H. $2t + 2$
I. $\frac{t}{4}$

8. The expression $\frac{n(n-1)}{2}$ is used to find the total number of different lines that can connect $n$ number of points. If there are 8 points, what is the total number of different lines that can connect them?

9. Windee owes Lanesha $10. Nell owes Windee $12. Lanesha owes Nell $14. What one payment will settle all debts?

   A. Nell pays Windee $12.
   B. Lanesha pays Nell $14.
   C. Lanesha pays Windee $10.
   D. Windee pays Lanesha $12.

10. Students conducted an experiment in Mrs. Sweeterman's sixth-grade class. The students measured the temperature increase of water heated gradually in a pot. One student recorded the temperatures in the table below.

| Number of Minutes | Degrees (Fahrenheit) |
|---|---|
| 2 | 46 |
| 4 | 70 |
| 6 | 94 |
| 8 | 118 |
| 10 | 142 |
| $t$ | |

If $t$ represents time passed in minutes, which expression can be used to find the water temperature after $t$ minutes?

   F. $10t + 22$
   G. $12t + 22$
   H. $-22 - 10t$
   I. $-22 - 12t$

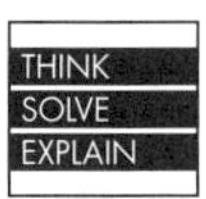


11. A landscape wholesaler supplies flowers to local nurseries. The wholesaler charges $1.29 per carton of flowers.

Part A. Write a rule or expression to determine the cost of $n$ number of cartons of flowers.

______________________________

______________________________

______________________________

Part B. Use the rule or equation you wrote to find the cost of 24 cartons of flowers.

12. If $y = \frac{16 + x}{4}$, what is the value of $y$ when $x = 20$?

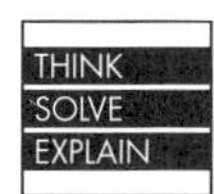


13. Crickets chirp faster in warm weather. It is possible to use formulas to calculate the number of times a cricket will chirp per minute.

Part A: The formula used for converting from degrees Celsius to degrees Fahrenheit is

$F = \frac{9}{4}C + 32,$

where $F$ = temperature in degrees Fahrenheit and $C$ = temperature in degrees Celsius. If the temperature is 24 degrees Celsius, what is the temperature in degrees Fahrenheit? Show or explain how you got your answer.

Part B: The number of chirps per minute $c$ a cricket makes is based on the temperature in degrees Fahrenheit $F$. Use the equation $c = 4F - 160$ to find the number of chirps per minute when the temperature is 24 degrees *Celsius*. Use your answer from A, and show or explain how you got your answer to B.

**14.** Which equation means "A number decreased by four is fifty-two?

**A.** $4 - n = 52$ **C.** $n = 4 + 52$
**B.** $n + 4 = 52$ **D.** $n - 4 = 52$

**15.** The weekly cost C in dollars of producing boat console covers at Kanvas Kovers is given by the expression

$$C = 5000 + 500x,$$

where $x$ is the number of console covers produced. How many console covers did the company produce in a week when the cost was $25,000?

## ANSWERS TO SAMPLE QUESTIONS

1. **C.** Although A works for the first $(x, y)$ pair, it does not work for all pairs. Substitute at least two pairs into the equations to check your answer.

$$y = 3x - 10$$
$$-4 = 3(2) - 10$$
$$-4 = 6 - 10$$
$$-4 = -4$$

Since the two sides are equal, C is the correct equation.

2. **F.** The open circle indicates $a <$, while the closed circle indicates $a \leq$. For a compound inequality, numbers must be arranged left to right from lowest to highest.

3. Part A. $f + 9 < 23$

Part B.

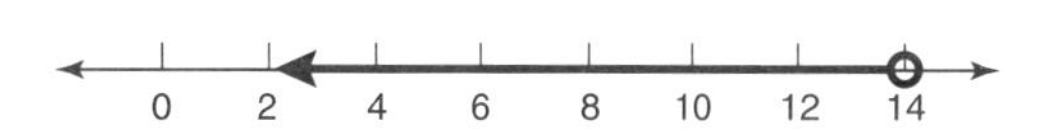


4. **−32.** Substitute −3 for $x$ in the equation: $y = 9(-3) - 5$; $y = -27 - 5 = -32$

5. **75.** Substitute 600 for distance $d$ and 8 for time $t$ in the formula and solve: $600 = 8r$. Divide both sides by 8: $r = 75$.

6. **B.** Height $h$ in inches can be found by multiplying 5 by the horizontal length. Since the length is 15 ft, the height is found by multiplying 5 in by the number of feet in the length: $5 \cdot 15 = 75$ in. Divide by 12 to convert inches to feet: 75 in ÷ 12 = 6 ft, 3 in.

7. **H.** Multiply the number of tables by 2 and add 2: $2t + 2$. If you have difficulty in recognizing the pattern, evaluate each expression by substituting the number of tables into the expression and checking to see whether the answer matches the number of seats.

8. **28.** Substitute 8 for $n$:

$$\frac{8(8-1)}{2} = \frac{8(7)}{2} = \frac{56}{2} = 28.$$

9. **B.** If Lanesha pays Nell \$14, then Nell will be able to repay the \$12 she owes Windee. In turn, Windee will now have the \$10 she needs to pay Lanesha.

10. **G.** Multiply the number of minutes by 12 and add 22 to get the temperature.

11. Part A. **$1.29c$**

    Part B. $1.29 \times 24 =$ **\$30.96**

12. **9.** Substitute 20 for $x$ in the equation:

$$y = \frac{16+20}{4} = \frac{36}{4} = 9.$$

13. Part A. **86**

$F = \frac{9}{4}C + 32$
$F = 9 \times 24 \div 4 + 32$
$F = 216 \div 4 + 32$
$F = 54 + 32$
$F = 86°$

Part B. **186**

$c = 4F - 160$
$c = 4(86) - 160$
$c = 344 - 160$
Chirps per minute = 186

14. **D.**

15. **40**

| | |
|---|---|
| $C = 5000 + 500x$ | Write the formula. |
| $25{,}000 = 5000 + 500x$ | Substitute 25,000 for C. |
| $20{,}000 = 500x$ | Subtract 5000 from both sides. |
| $40 = x$ | Divide by 500. |

# Lesson 19

# Reading and Interpreting Graphs

## BAR GRAPHS

The bar graph shows the results of a math test given to 60 students. Find the difference between the number of students who received A or B and the number who received C, D, or F.

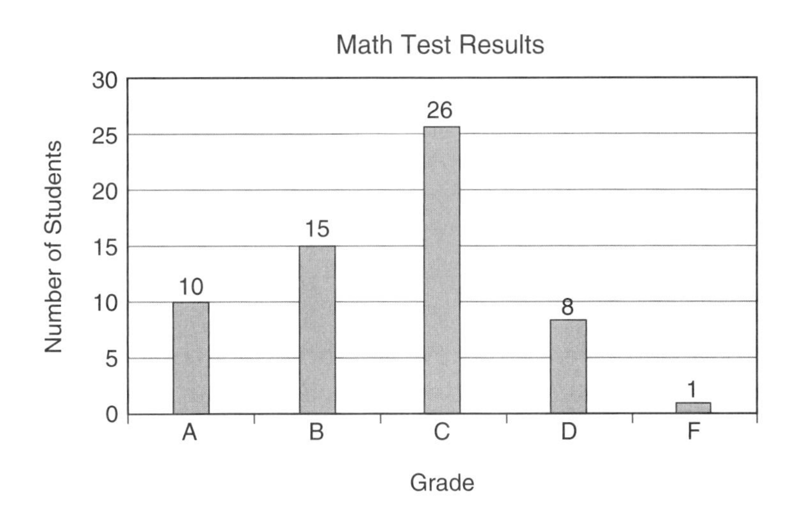


The number of students who received A or B is 25 (10 + 15). The number of students who received C, D, or F is 35 (26 + 8 + 1). The difference between the numbers is found by subtracting: 35 – 25 = 10.

In a *bar graph*, each bar represents data. Bar graphs are used to *compare* data. The bar graph shown compares the grades of students who took a math test. The vertical scale tells how many students received each grade.

Notice that the graph has a title and that the vertical and horizontal axes are labeled so that the reader has a clear idea of what the graph shows. When you design a bar graph, decide on the labels for the axes before beginning. Always label the graph's axes, and give your graph a title.

Each bar should be exactly the same width, with equal spaces (intervals) between the bars. As you draw a bar graph, give a lot of thought to any scales you use along the axis.

In this bar graph, the interval on the vertical axis is 5. You may count by any number, such as 2, 5, 10, or 100 (the most common), as long as all the numbers will "fit" into the grid. It is not a good idea to add lines to the grid you are given on the FCAT, so check ahead of time to make certain that the scale you choose will fit.

The graph on page 173 can be also organized into a horizontal bar graph as shown below. Notice that the labels are reversed.

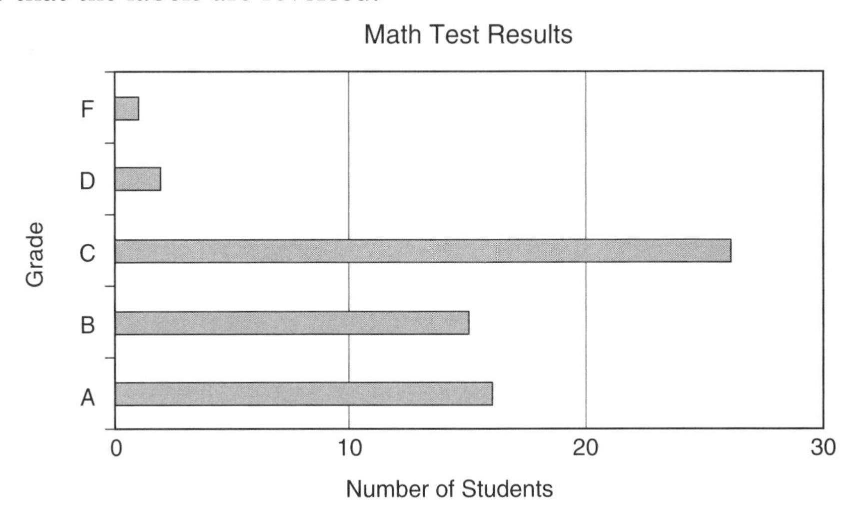


## DOUBLE-BAR GRAPHS

Double-bar graphs compare two **similar** sets of data. In the double-bar graph shown here, high and average wind speeds for several U.S. cities are compared. A double-bar graph should always have a *legend* or key that shows the meaning of each bar color or shade. Notice that the intervals between related pairs of bars are equal.

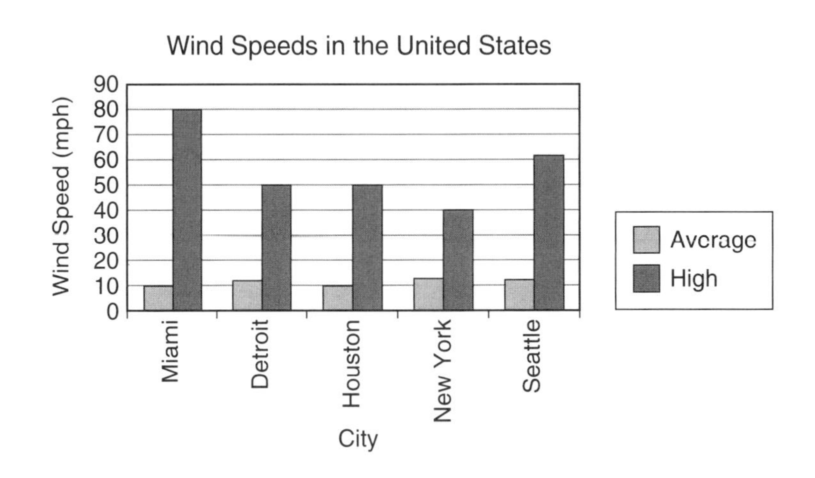

## HISTOGRAMS

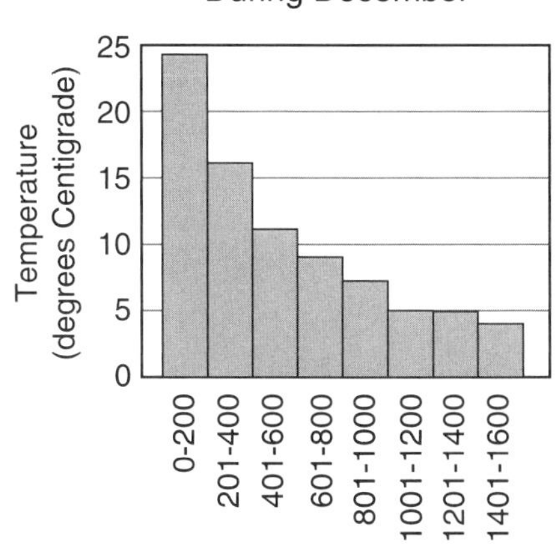


*Histograms* are similar to bar graphs in that each bar represents data. However, in a histogram the data are grouped in intervals, such as 0–200 and 201–400. Because the data are grouped, there is usually no space between the bars. The grouped data must be expressed in equal intervals. For example, 1980–1984 represents an interval of 5 years; therefore, the next intervals should be 1985–1989 and 1990–1994.

## STEM-AND-LEAF PLOTS

A *stem-and-leaf plot* is a method of organizing numerical data that allows you to see how the data fall into intervals while still showing every number in the set. The leaf portion of the plot is always just one digit, representing the lowest place value, and the stem is the higher place value. The stem can be more than one digit.

**Example 1:** Ms. Bedford wanted to see how her class performed on the FCAT pretest. Their scores were as shown in the stem-and-leaf plot. Find the range, mean, median, and mode of this data group.

| Stem | Leaf |
|---|---|
| 30 | 3 |
| 31 | 0 1 3 4 4 5 6 7 9 9 9 |
| 32 | 3 3 7 |
| 33 | 1 2 6 6 |
| 34 | 8 |

Key: 32|3 = 323

The *range* of the scores is found by subtracting the lowest score from the highest: 348 – 303 = 45.

The *mean* is the average of all the scores. Add the scores together, and divide by the number of scores: 6426 ÷ 20 = 321.3.

The *median* is the score in the middle. If there is an even number of scores, add the two middle scores together and divide by 2: (319 + 319) ÷ 2 = 319.

The *mode* represents the score seen most often. According to the stem-and-leaf plot, 319 occurs more often than any other score.

**Example 2:** The stem-and-leaf plot represents the temperatures recorded in Miami during 2 weeks in the month of January: 65, 66, 73, 75, 70, 67, 68, 82, 80, 85, 74, 76, 70, 71.

| Stem | Leaf |
|---|---|
| 6 | 5 6 7 8 |
| 7 | 0 0 1 3 4 5 6 |
| 8 | 0 2 5 |

Key: 8|0 = 80

Notice that the stem for these data represents the tens place and that the leaf represents the ones place. For each stem, the leaves are in order from least to greatest.

## LINE GRAPHS

*Line graphs* show change over time. The double line graph below compares the change in math SAT scores of college-bound students over the period 1995–2000. What conclusions can you draw about the comparison between male and female SAT scores shown in the graph?

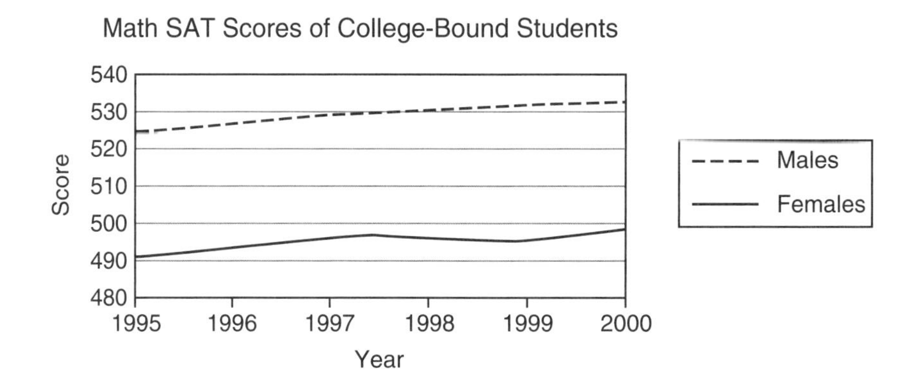

Males scored higher overall than females. In fact, males scored about 35 points higher. Both male and female SAT scores showed a rise over time.

## SCATTERPLOTS

A *scatterplot* is a group of points showing a relationship between two sets of data. When the points form a pattern, it is called a *trend.* A scatterplot can have a positive trend, a negative trend, or no trend.

### Positive Trend

When the data forming a scatterplot rise from left to right, the scatterplot is said to have a *positive trend.* In other words, one set of data rises as the other rises. For example, you would expect test scores to increase as study time increases.

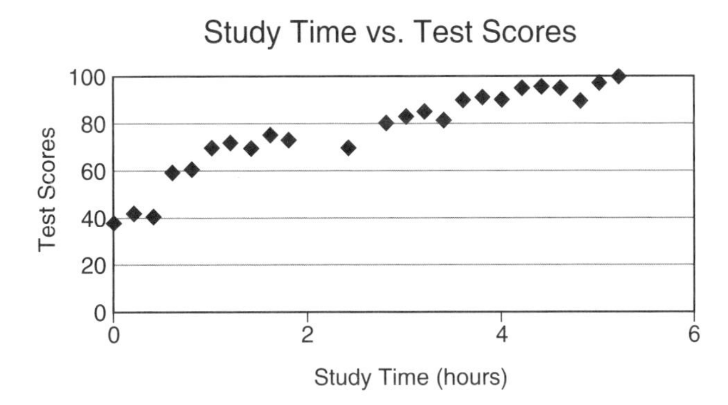


### Negative Trend

If the data forming a scatterplot go down from left to right, the scatterplot has a *negative trend.* One set of data rises, while the other decreases. In the graph shown here, the cost per ounce of cereal is compared to the size of the box. As you can see, the larger the box of cereal, the cheaper the cereal is per ounce.

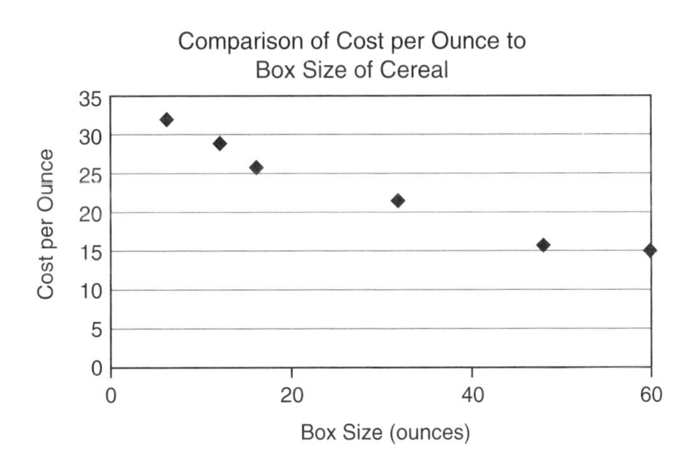

## No Trend

A scatterplot with *no trend* has a scattering of points that take no specific direction. The two sets of data have *no relationship* to each other. Notice in the scatterplot shown here that there is no relationship between a person's height and his or her intelligence (IQ).

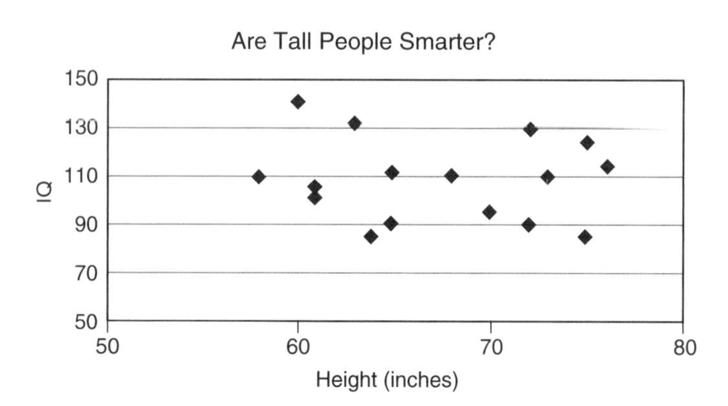


## Trend Lines

A *trend line* is a line that, when drawn through a scatterplot, travels in the same general direction as the data points and passes through the center of the data. In studying for the FCAT, all trend lines tested are straight lines. Trend lines are very handy for predicting future values. When the data have a strong positive or strong negative trend, the trend line fits the data very closely with few, if any, points wandering away from the line. A prediction made from such a trend line tends to be more accurate than one based on a trend line in which the data points stray away from the line.

There is a lot of talk these days about global warming. The graph shown contains data about average global temperatures. You could predict that temperatures will continue to rise by extending the trend line (the solid line on the graph) into future years. You might also predict that the average global temperature will be around 58 degrees by the year 2030.

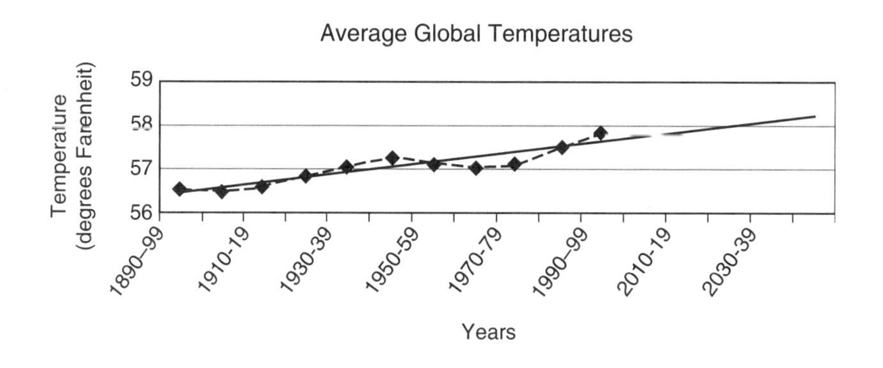

# CIRCLE GRAPHS

A *circle graph* can show how a whole is broken into parts. Frequently, circle graphs are used to show percentages. On the FCAT, you may be asked to display data in a circle graph.

**Example:** Ann's monthly budget is broken down as follows: 42% for rent, 21% for car payments, 16% for food, 7% for entertainment, and 14% for miscellaneous. Is the graph shown drawn correctly? Why or why not?

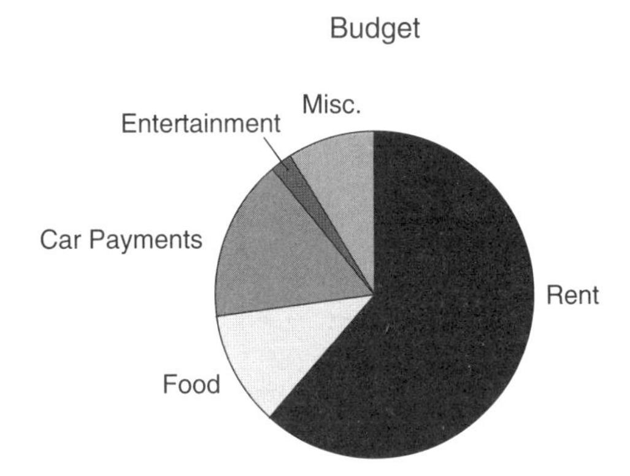


The graph is drawn incorrectly. For example, if rent is 42%, that item should represent less than half of the graph. Instead, rent takes up more than half. Also, it is customary to show each percent or value directly on the graph.

If the graph were drawn correctly, it would look as shown here.

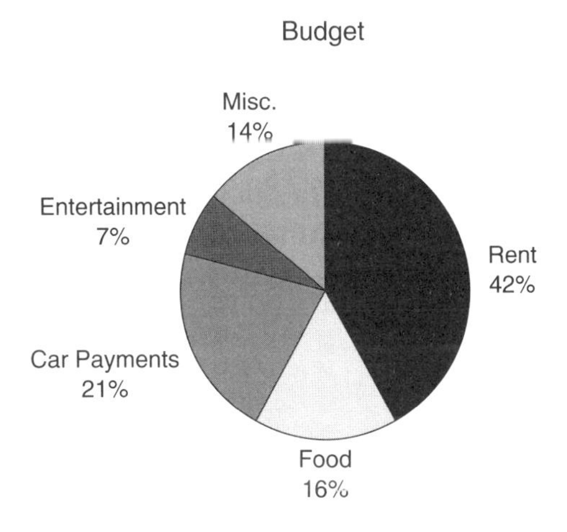

## VENN DIAGRAMS

A *Venn diagram* is an additional way to organize and display data. Venn diagrams organize sets of numbers or things. The sets can then be compared to see what they have in common.

In the accompanying Venn diagram, Mr. Jones has recorded the number of students wearing shorts and the number of students wearing tennis shoes in his class. There are 35 students in the class.

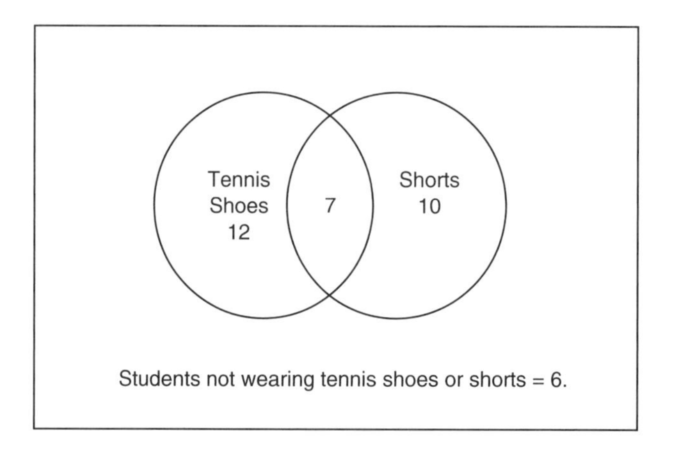


The following observations can be made from this Venn diagram:

1. The total number of students wearing tennis shoes is 19. Count *all* the students in the circle containing the words *Tennis Shoes.*
2. The total number of students wearing shorts is 17. Count *all* the students in the circle containing the word *Shorts.*
3. The number of students wearing *both* tennis shoes and shorts is 7. Notice that the circles *share* these 7 students. This is the place where the two sets *intersect.*
4. The number of students wearing tennis shoes but not shorts is 12.
5. The number of students wearing shorts but not tennis shoes is 10.
6. The total number of students in Mr. Jones' class is 35. Add: 12 + 7 + 10 + 6 = 35.

# SAMPLE QUESTIONS

## Reading and Interpreting Graphs

1. To analyze the high school graduation rate by state, Lennie used the accompanying table to draw a graph. Which graph represents the best way to graph these data?

| State | Percent Graduating |
|---|---|
| Colorado | 72 |
| Florida | 57 |
| Louisiana | 50 |
| Missouri | 72 |
| New York | 62 |
| Oregon | 68 |

A.

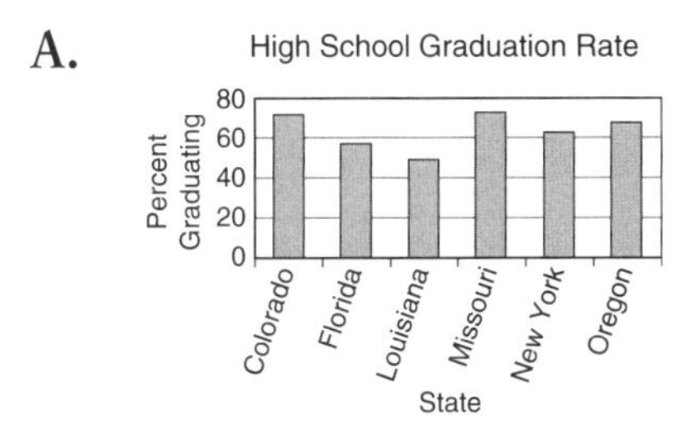


B.

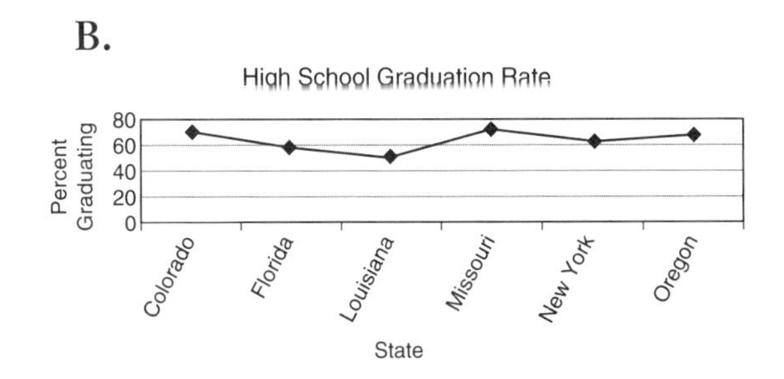


C.

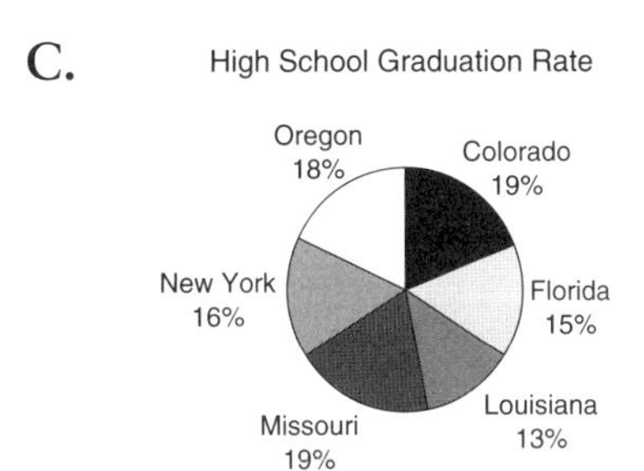


D.

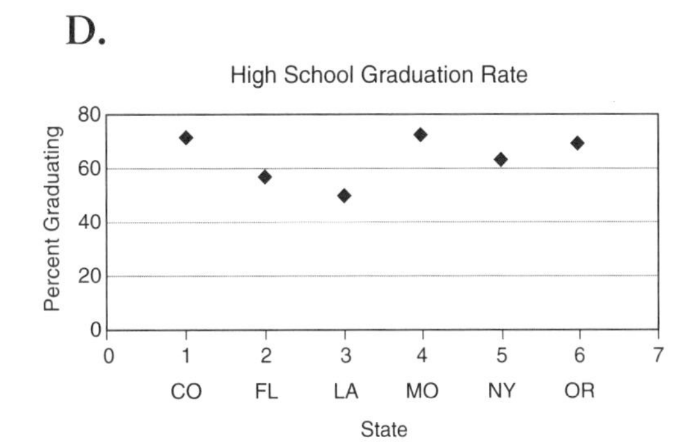


2. This stem-and-leaf plot represents the temperatures recorded in Tampa during the month of March.

| Stem | Leaf |
|---|---|
| 6 | 5 6 7 |
| 7 | 0 0 1 3 4 5 6 8 |
| 8 | 0 2 5 |

On how many days did the temperature rise above 70 degrees?

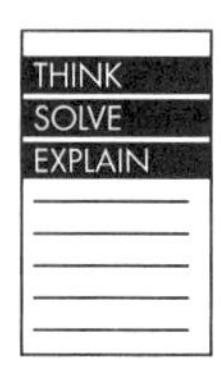


3. The accompanying graph shows data collected during a class exercise. The question posed by the teacher was "Is there a relationship between height and shoe size?" Students gathered the data by measuring each other's height and foot length.

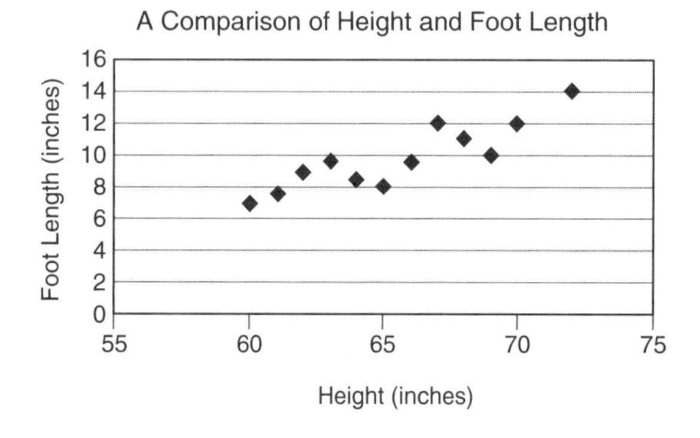


Part A. On the graph, draw a trend line for the data.

Part B. Predict, using your trend line, the approximate foot length for a person 75 inches tall.

Part C. Complete the following statement: The taller the person, the ________ the shoe size.

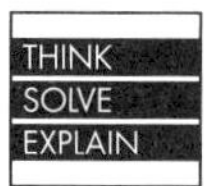


4. The table shows population changes, in millions, in several Florida counties from 1990 to 2000.

| County | Population 1990 | Population 2000 |
|---|---|---|
| Duval | 680 | 730 |
| Pinellas | 860 | 910 |
| Hillsborough | 830 | 960 |
| Polk | 860 | 890 |
| Orange | 690 | 820 |

Part A. On the grid, make and label a double-bar graph for the data given.

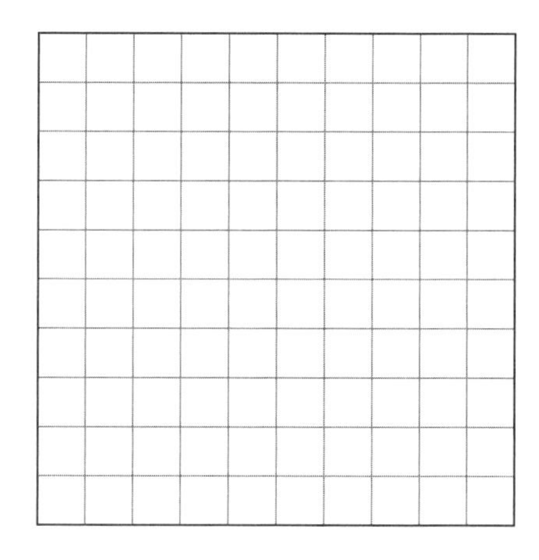

Part B: Which county showed the most growth during this time period? Which county showed the least growth? Explain the difference.

______________________________

______________________________

______________________________

Mr. Harris's music class surveyed 40 students to determine their favorite type of music. The results are shown in the Venn diagram. Use it to answer questions 5 and 6.

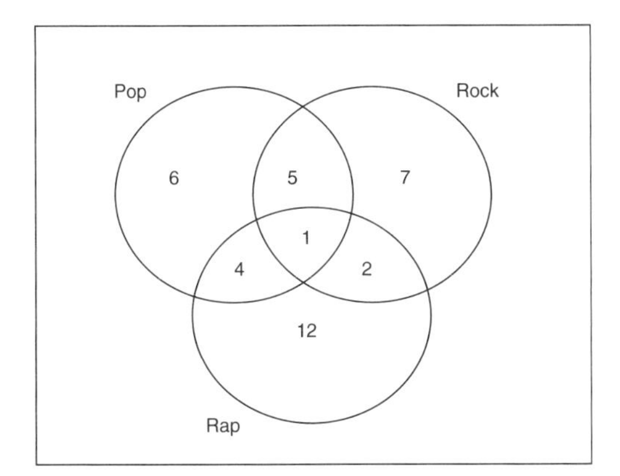


5. How many students like pop and rock but not rap?

6. How many students like pop, rock, and rap?

7. After visiting an amusement park, Steven plotted time (minutes aloft) versus height above ground on a Ferris wheel. Which graph could Steven have drawn?

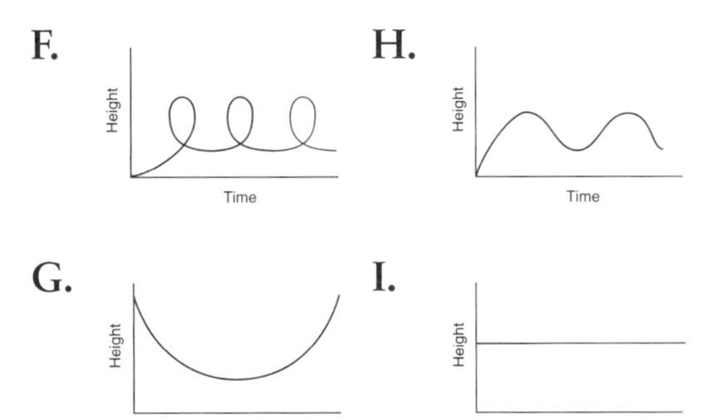


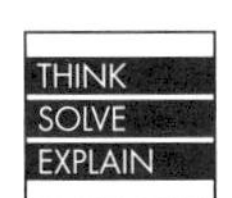


8. Kyle raises tomatoes in a greenhouse. In August, on the advice of the local nursery, Kyle changed the plant food he was feeding his tomato plants.

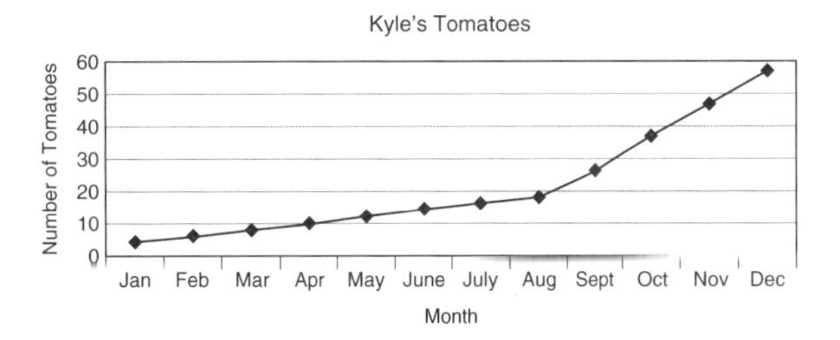


Describe the changes that took place as a result of the new plant food.

________________________________

________________________________

________________________________

9. Pam had paint mixed at the paint store. The following pigments were added to a base color to make the beige she wanted for her walls: 2 parts lamp black, 1 part yellow, 1 part green, and 1 part red. If Pam were to graph the pigments used, which type of graph would best show these portions?

   A. bar graph
   B. line graph
   C. circle graph
   D. Venn diagram

10. Mr. Harris has two bank accounts. He followed their balances over a period of 10 weeks. During which of the 10 weeks was the value of the combined accounts the greatest?

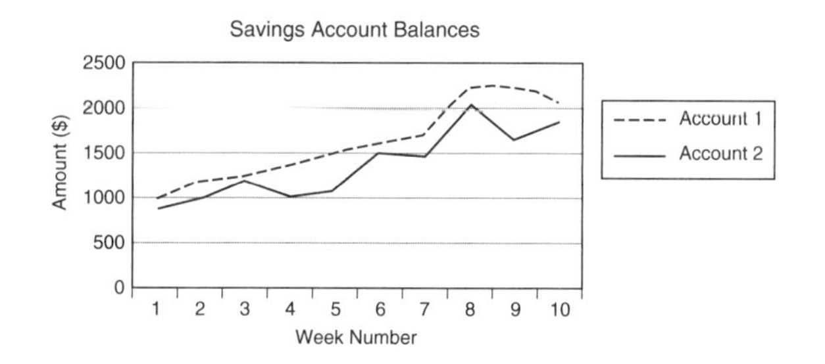


   F. 3
   G. 6
   H. 7
   I. 8

# ANSWERS TO SAMPLE QUESTIONS

1. **A.** Bar graphs are used to compare data. Here, the data represent the percent of students graduating in each state.

2. **9.** The temperatures above 70 are 71, 73, 74, 75, 76, 78, 80, 82, and 85.

3. Part A.

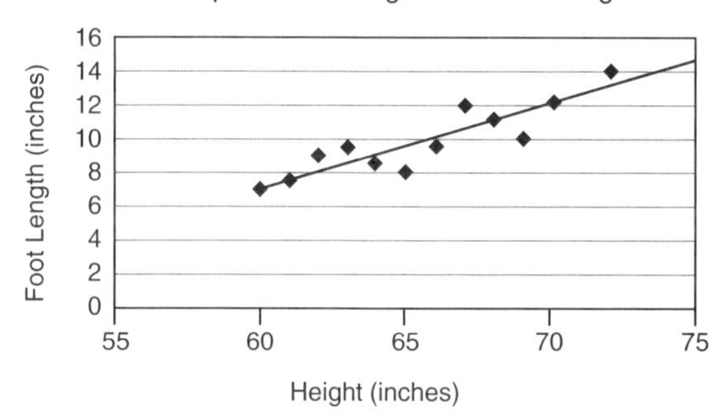


Part B. **14.5 in.**
Part C. The taller the person, the **bigger** the shoe size.

4. Part A.

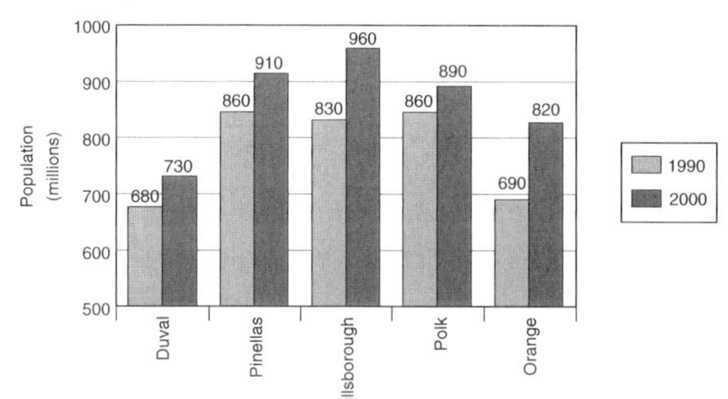


Part B: **Hillsborough County** showed the most growth.
**Polk County** showed the least growth.

The bars that are furthest apart indicate the most growth, while the bars closest together show the least growth.

5. **5.** The area where the pop and rock circles intersect shows students who like these types of music.

6. **1.** Only one student likes all three: pop, rock, and rap.

7. **H.** Only graph H shows that the Ferris wheel begins at ground level, goes higher with time, and then goes lower. If you examine graph F closely, it appears to actually go backward in time. Graph G starts above the ground and then descends, while graph I never changes height at all.

8. **The plant food increased the number of tomatoes that were produced.**

9. **C.** A circle graph would best show the portions.

10. **I.** The value of the combined accounts was at a maximum during week 8.

# Lesson 20

## Probability, Odds, and Data Analysis

If the prediction of rain today is 70%, should you plan on taking your umbrella?

Yes, because 70% is much closer to 100% than to 0%, making the event of rain very likely, although not guaranteed.

### PROBABILITY

The term *probability* helps you to determine how often something is likely to happen. When the weather channel forecasts a 70% chance of rain, you know that the probability of rain is $\frac{70}{100}$, $\frac{7}{10}$, or 0.7. You cannot be 100% sure that it will rain, or exactly when it will rain, but you can use this information to predict the likelihood of rain.

The probability of an event is usually expressed as a number between 0 and 1. This number can be written as a fraction, decimal, or percent. A probability of 0 means the event will *never* happen; a probability of 1 means the event will *always* happen.

**Example 1:** The probability of getting heads when you flip a fair coin is $\frac{1}{2}$.

Therefore, it is equally likely that you will get heads or tails because $\frac{1}{2}$ is just as close to 0 as it is to 1.

**Example 2:** The probability of drawing a black marble out of a jar is given as 0.28. Therefore, you are more likely to draw some other color because 0.28 is closer to 0 than it is to 1.

If you need to work out the probability that something will happen (an event), decide on the total number of things that can happen. Each of these is called an *outcome*. For example, if you throw a fair six-sided die (one of two or more dice), there are six possible outcomes: {1, 2, 3, 4, 5, 6}. The probability that you will roll any particular number is:

$$P(\text{event}) = \frac{\text{number of favorable outcomes}}{\text{total possible outcomes}}.$$

**Example 3:** The probability that you will roll a 3 is given as $P(3) = \frac{1}{6}$. There is only one 3 on a die, so there is only 1 favorable outcome out of 6 possible outcomes.

**Example 4:** The probability that you will roll an even number is given as

$$P(\text{even}) = \frac{3}{6} = \frac{1}{2}.$$

Because there are 3 even numbers, there are 3 favorable outcomes out of 6 possible outcomes.

**Example 5:** The probability that you will roll a number divisible by 3 is given as

$$P(\text{divisible by } 3) = \frac{2}{6} = \frac{1}{3}.$$

There are 2 numbers that are divisible by 3 (3 and 6), so there are 2 favorable outcomes out of 6 possible outcomes.

**Example 6:** If you roll an 8-sided die (numbered from 1 to 8), what is the probability of rolling a number divisible by 4?
What is the probability of rolling a number that is NOT divisible by 4?

The possible outcomes are {1, 2, 3, 4, 5, 6, 7, 8}. Only 2 of these numbers are divisible by 4: {4, 8}. The other numbers, {1, 2, 3, 5, 6, 7}, are not divisible by 4. The probability you will roll a 4 or an 8 is given by

$$P(\text{divisible by } 4) = \frac{2}{8} = \frac{1}{4}.$$

The probability that you will roll a number that is NOT divisible by 4 is given by

$$P(\text{not divisible by } 4) = \frac{6}{8} = \frac{3}{4}.$$

The probability that something WILL happen plus the probability that it WILL NOT happen equals 1 (or is 100%).

## Theoretical Probability Versus Experimental Probability

If you roll a six-sided die, the chances of rolling any particular number is $\frac{1}{6}$. If you roll this same die 12 times, you would *expect* that each number on the die would come up exactly twice. The expected probability is called *theoretical probability*. You conduct an *experiment* when you actually roll a die 6 times and keep track of how many times you get 1, 2, 3, 4, 5, or 6.

**Example:** Cindy conducts an experiment by rolling a six-sided die 12 times. In the table, the theoretical or expected results are shown and compared to the actual or experimental results.

| Number on Die | Theoretical Results | Experimental Results |
|---|---|---|
| 1 | 2 | 1 |
| 2 | 2 | 3 |
| 3 | 2 | 0 |
| 4 | 2 | 3 |
| 5 | 2 | 2 |
| 6 | 2 | 3 |

*Experimental probability* generally comes close to theoretical probability, but is frequently not exactly the same. This is especially true when only a few trials are conducted. Remember that theoretical probability is what is *mathematically* expected, but the prediction is based on many more experiments than would normally be conducted. If Cindy had rolled the die one thousand times, she could expect that her results would be closer to the theoretical results than they were in only 12 tries.

## ODDS

*Odds* are another form of probability. The odds that you will throw heads on the toss of a fair coin are 1 to 1, or even, because only two things can happen: one is heads, and the other is tails. Odds are most often expressed as a fraction or ratio.

On the toss of a fair six-sided die, what are the odds that you will roll a 4?

What are the odds that you will NOT roll a 4?

The odds of rolling a 4 are 1:5. There is only 1 favorable outcome, rolling a 4. There are 5 unfavorable outcomes, rolling a 1, 2, 3, 5, or 6.

The odds of NOT rolling a 4 are 5:1.

**Example 1:** Burns Middle School is playing soccer against Dowdell Middle School. The odds of a Burns win are given as 3 to 2, or $\frac{3}{2}$. From these odds, you know that Burns Middle is expected to win. Odds of 3 to 2 mean that, if the teams play 5 games, Burns Middle should win 3 and Dowdell Middle should win 2. The odds that Burns Middle will lose are given as 2 to 3.

**Example 2:** A basketball player makes a point 1 out of every 3 times he shoots. If he has to make the winning shot, what are the odds that his team will win?

The odds that he will make the winning shot are 1:2 because, out of every 3 shots, he makes 1 and misses 2.

# DATA ANALYSIS—RANGE, MEAN, MEDIAN, AND MODE

## Range

The test scores in Miss Care's social studies class were 100, 65, 95, 80, 70, and 90. What was the range of scores?

The range was 35. Range is found by taking the highest score (100) and subtracting the lowest score (65).

## Mean

The *mean* is a measure of central tendency, usually referred to as the *average*. It is calculated by adding all the data points and dividing by the number of data points given. The mean of the social studies scores for Miss Care's class is found this way:

$$\frac{100+65+95+80+70+90}{6} = 83.3.$$

## Median

The *median* is another measure of central tendency. If you put a data set in order from least to greatest (or greatest to least), the *median* is the number that falls exactly in the middle. If the data set contains an odd number of data points, the median will be the middle number. If there is an even number of data points in the set, the median will be the mean of the two middle numbers.

**Example 1:** Find the median of the following data set: {2, 7, 4, 7, 5}.

Step 1: Put the numbers in order from least to greatest: {2, 4, 5, 7, 7}.

Step 2: Locate the number in the middle: {2, 4, <u>5</u>, 7, 7}. That number is the median.

**Example 2:** Find the median of the data in the stem-and-leaf diagram.

| Stem | Leaf |
|---|---|
| 1 | 0 1 1 2 4 9 |
| 2 | 1, 3, 5, 7 |
| 3 | 5, 6, 6, 8 |

The numbers in the stem-and-leaf diagram have already been placed in order. There are two middle numbers {21, 23}. Add them and then divide by 2: (21 + 23) ÷ 2 = 22.

## Mode

The third measure of central tendency is the mode. The *mode* describes the value in a data set that occurs most often. For example, in the data set {5, 6, 3, 4, 2, 4, 7, 4}, the 4 occurs most often. The mode is a good choice for a measure of central tendency if a data set contains a lot of numbers that are identical.

A data set can have *no mode*, for example {1, 2, 3, 5}; have *one mode*, {1, 2, 3, 3, 5}; or be *bimodal* (have *two modes*), {1, 1, 2, 2, 3, 4, 5, 6}.

# BOX-AND-WHISKERS PLOTS

A box-and-whiskers plot is a good way to display large sets of data. This method groups the data into four equal parts called *quartiles*. If you have 100 data values, 25 of them will be in the lower one-fourth of the plot, 25 in the upper one-fourth of the plot, and the remaining one-half will be in the "box."

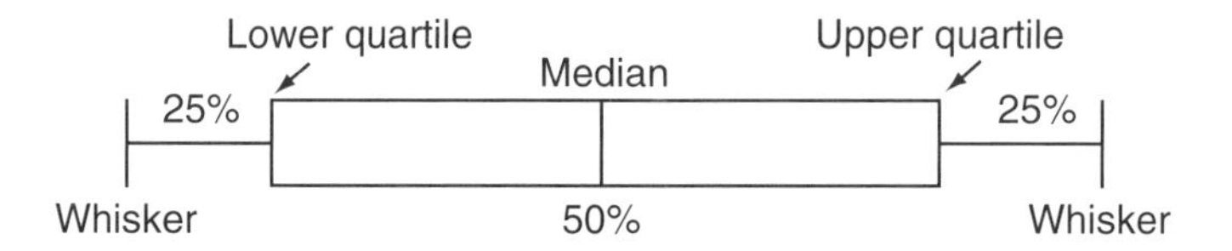


A box-and-whiskers plot is constructed using five numbers:

- The lowest number (the left whisker).
- The highest number (the right whisker).
- The median of the data group.
- The lower quartile (the median of the lower half of the numbers).
- The upper quartile (the median of the upper half of the numbers).

### Example 1:

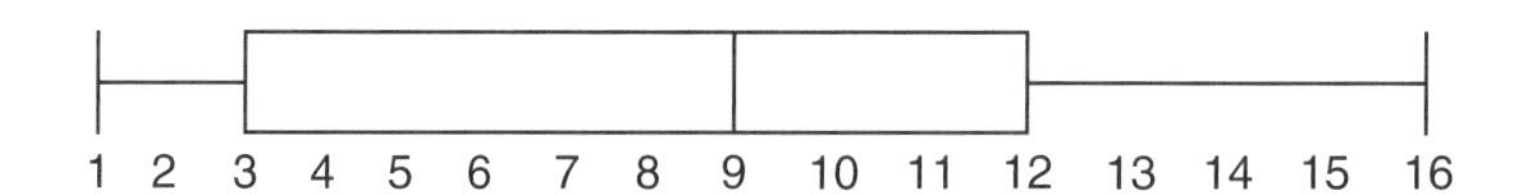


- Lowest number (also called *lower extreme*) = 1
- Highest number (also called *upper extreme*) = 16
- Median = 9
- Lower quartile = 3
- Upper quartile = 12

The range is 16 – 1 = 15.

- 50% of the data is between 3 and 12.
- 25% of the data is between 1 and 3.
- 25% of the data is between 12 and 16.
- 50% of the data is above 9.
- 50% of the data is below 9.

# SAMPLE QUESTIONS

## Probability, Odds, and Data Analysis

1. Some students are making tetrahedron kites out of tissue paper in their math class. To decide which color each student will use, the names of the colors are written on pieces of paper and placed in a box.

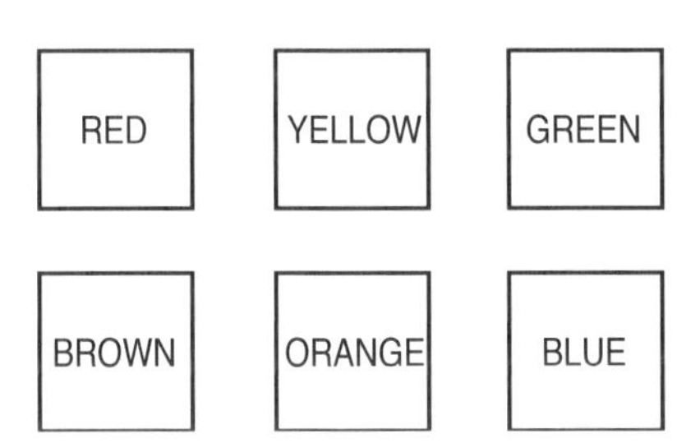


Each student reaches into the box and selects a slip of paper without looking. If Louis goes first, what is the probability that he chooses either green or blue?

A. $\frac{1}{6}$

B. $\frac{1}{2}$

C. $\frac{1}{3}$

D. $\frac{2}{3}$

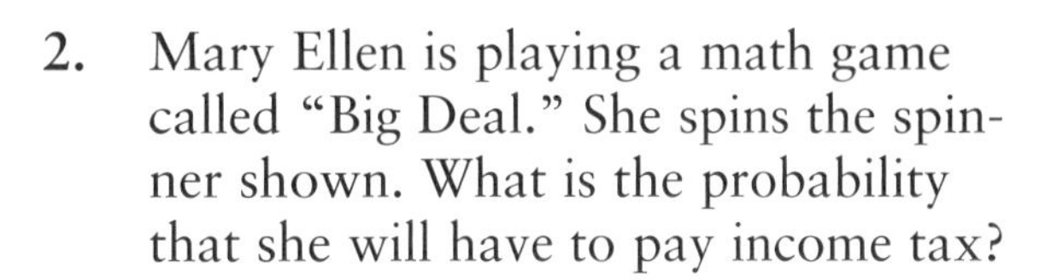

2. Mary Ellen is playing a math game called "Big Deal." She spins the spinner shown. What is the probability that she will have to pay income tax?

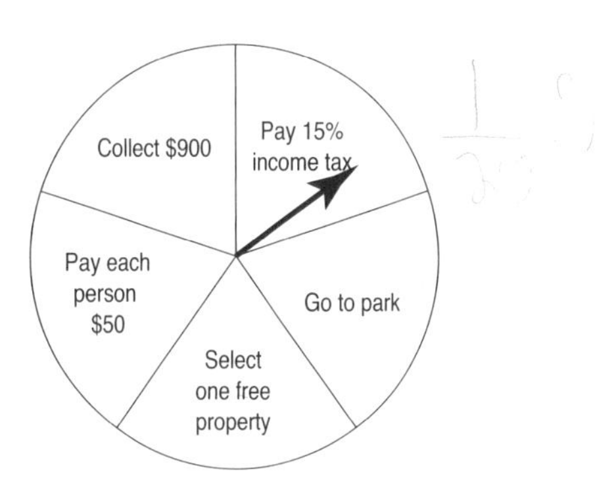


3. Irena, Raven, Mike, and Chad are seated in the front row of the movies. If they sit in the first four seats, in how many *different* orders can they seat themselves?

4. Mrs. Wilson needs to find the mean of her class's test scores. Which of the following procedures should she follow?

F. Arrange the scores in order, and select the middle score.

G. Add the scores, and divide by the number of tests.

H. Find the test score that the most students received.

I. Subtract the lowest score from the highest score.

5. Bud counted 58 M&Ms in his bag of candy. The bar graph shown represents the number of each color of M&Ms in Bud's bag. Which color represents the mode?

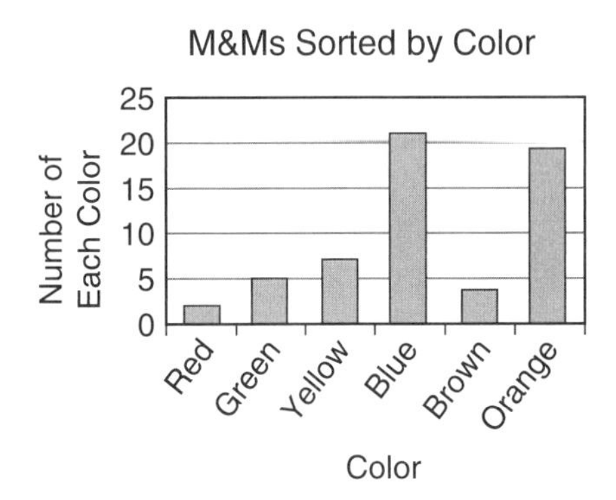


A. red
B. yellow
C. orange
D. blue

6. The box-and-whiskers plot shown represents the hourly wage paid each employee at Sander's Quickee Mart.

Which of the following best represents the median hourly wage?

F. \$5.00
G. \$10.50
H. \$8.25
I. \$13.00

7. The stem-and-leaf chart shows the ages of people who registered to receive a free turkey at a grocery store give-away.

| Stem | Leaf |
|---|---|
| 2 | 0, 2, 2, 4, 8, 9, 9 |
| 3 | 1, 3, 5, 7, 8 |
| 4 | 2, 3, 4, 4, 5 |
| 5 | 0, 2, 4, 5, 7, 7, 8 |
| 6 | 1, 1, 6, 9 |

One registration card was drawn at random. What is the probability the winner was 61 years old?

A. $\frac{1}{2}$
B. $\frac{1}{4}$
C. $\frac{1}{8}$
D. $\frac{1}{14}$

8. The temperatures taken at 1:00 P.M. in Tampa for 10 days in June are shown below:

92°, 89°, 84°, 96°, 92°, 88°, 89°, 97°, 90°, 98°

What was the median temperature for those ten days?

F. 89°
G. 90°
H. 91°
I. 92°

9. Sal is working with polyhedron dice. He selects a die that has nine sides, numbered from 0 to 8. If he rolls the die, what are the odds that he will roll a number divisible by 3?

A. 2:7
B. 2:8
C. 2:9
D. 7:2

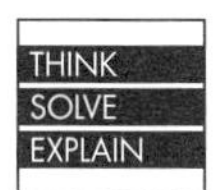


10. A farmer purchased 150,000 seeds to plant. The supplier told him that, of 150,000 seeds, 135,000 seeds are expected to germinate.

Part A: What percent of seeds are predicted to germinate?

Part B: Of the 150,000 seeds purchased, 145,000 seeds actually germinated. What percent germinated?

Part C. What possible reason could explain why the actual germination rate was better than the rate predicted? Explain, using what you know about theoretical and experimental probability.

______________________________

______________________________

______________________________

11. Angie wanted to find out whether students in her school liked math. She did a survey by going to the Math Club meeting after school and asking the 16 members how many liked math. Was this a valid survey? Choose the correct answer(s) from the list below.

I. The survey was valid. Every student who liked math had a chance to participate.
II. The survey was not valid; she should have given every student in the school a chance to participate in the survey.
III. The survey was not valid; she didn't ask enough people.
IV. The survey was not valid; she asked only people who were interested in math.

**F.** I only
**G.** II, III, and IV only
**H.** II and III only
**I.** II and IV only

## ANSWERS TO SAMPLE QUESTIONS

1. **C.** If Louis goes first, he has 2 chances out of 6, or a probability of $\frac{2}{6}$, which is $\frac{1}{3}$ in lowest terms, of choosing green or blue.

2. **$\frac{1}{5}$.** There are 5 equal parts in the circle. May Ellen has a 1 in 5 chance of selecting the income tax portion.

3. **24.** If Irena goes first, she has 4 seats to choose from. After Irena chooses, Raven has 3 seats to choose from. Once Raven is seated, Mike has 2 seats to choose from. After Mike is seated, Chad has to sit in the only remaining seat. Multiply $4 \times 3 \times 2 \times 1 = 24$ different orders.

4. **G.**

5. **D.** The graph shows that there are more blue M&Ms than any other color, so blue represents the mode.

6. **H.** The median in a box-and-whiskers plot is represented by the vertical line inside the "box."

7. **D.** Two people out of 28 were 61 years old. The probability of a 61-year-old person receiving the turkey is $\frac{2}{28}$, which is $\frac{1}{14}$ in lowest terms.

8. **H.** Arrange the numbers in ascending order: 84, 88, 89, 89, 90, 92, 92, 96, 97, 98. Find the middle number(s). There are two middle numbers (90 and 92); add them, and divide by 2 to get 91.

9. **A.** In the numbers 0 to 8, only two numbers are divisible by 3: 3 and 6. Therefore, 7 numbers are NOT divisible by 3. The odds of rolling a number divisible by 3 are 2:7.

10. Part A. **90.** The percent of seeds from this particular batch predicted to germinate is $\frac{135{,}000}{150{,}000} = 0.9 = 90\%$.

    Part B. **96.7.** $\frac{145{,}000}{150{,}000} = 0.966 = 96.7\%$.

    Part C. Theoretical probability describes the number of seeds predicted to germinate. The actual (experimental) number may be slightly more or less. In this case, the supplier probably sold seeds to other farmers; some of these seeds may have had lower germination rates.

11. **G.** A good survey should give everyone an equal opportunity to participate. By going only to the Math Club, Angie asked only people who had already shown an interest in math. Also, the Math Club was only a small part of the entire school population. Angie could have conducted a better survey by standing in front of the school as students entered in the morning and asking as many as she could.

# Practice Test 1

1. Audrey sent her daughter to the grocery store to buy sliced cheese. The daughter was told to get $\frac{2}{3}$ pound of mozzarella, $\frac{1}{4}$ pound of American, $\frac{1}{2}$ pound of Swiss, and $\frac{3}{4}$ pound of cheddar. Which of these packages of cheese weighs the most?

   **A.** mozzarella  **C.** Swiss
   **B.** American  **D.** cheddar

2. Sam is in charge of making sure the club pool is maintained with four thousand ninety-six gallons of water. What is another way to express this number?

   **F.** $7^8$  **H.** $256^{16}$
   **G.** $8^4$  **I.** $1024^4$

3. Evaluate $[(abc)^3]^4$.

   **A.** $a^7b^7c^7$
   **B.** $2(abc)^7$
   **C.** $a^{12}b^{12}c^{12}$
   **D.** $7a^27b^27c^2$

4. What is the result when the operation $\frac{\square}{\bigcirc} \times \frac{\bigcirc}{\triangle}$ is performed?

   **F.** $\frac{\square}{\triangle}$  **H.** $\frac{\bigcirc^2}{\square\triangle}$
   **G.** $\frac{\triangle\square}{\bigcirc^2}$  **I.** $\frac{\square\bigcirc}{\triangle}$

5. To simplify the expression $3 + 9 \div 2 \times 6 - 7$, which order of operations should be used?

   **A.** $+, \div, \times, -$  **C.** $\div, \times, +, -$
   **B.** $\times, \div, +, -$  **D.** $+, -, \times, \div$

6. Jack is 15 years old. He is 3 years older than $\frac{1}{3}$ of his mother's age. How old is Jack's mother?

   **F.** 35  **H.** 39
   **G.** 36  **I.** 54

7. Redee Concrete is pouring a rectangular driveway 12 feet by 30 feet for a homeowner. Suddenly, the owner of the house decides to double the dimensions of the driveway. What happens to the area of the concrete driveway?

   **A.** It increases by a factor of 2.
   **B.** It increases by a factor of 4.
   **C.** It increases by a factor of 8.
   **D.** It increases by a factor of 16.

8. Mary is on the girl's track team. She finds a map with a jogging trail that is marked into 13 equal intervals. The trail is 8.4 kilometers in length. If the map is drawn to scale, how long should each interval be?

F. 64 m  H. 646 m
G. 110 m  I. 6460 m

9. According to statistics, Gulf County in Florida encompasses an area of 559 square miles. If 17 million people live in Gulf County, what is the population per square mile?

A. 30  C. 3041
B. 304  D. 30,411

10. The diagram shows the net of a square pyramid. Which statement is true?

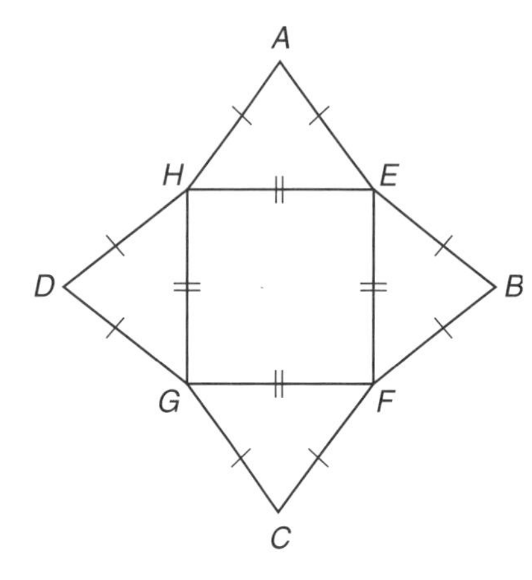


F. $\overline{GF} \cong \overline{GC}$
G. $\angle HEF \cong \angle EHA$
H. $\angle FGC \cong \angle GFC$
I. $\overline{BF} \cong \overline{FG}$

11. Quadrilateral *ABCD* is a parallelogram. What is the measure of $\angle CAD$?

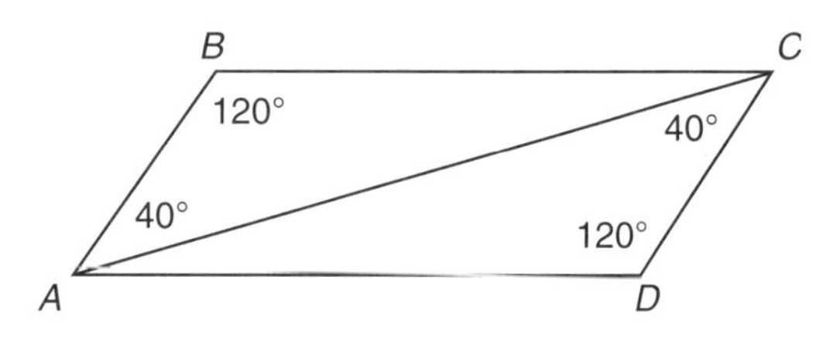


A. 120°  C. 40°
B. 60°  D. 20°

12. In the figure shown, the performance of which transformation(s) on one triangle will construct the tessellation?

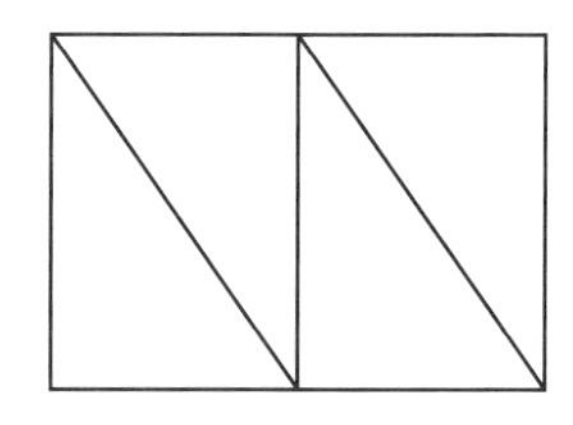

I. Translation
II. Reflection
III. Rotation

F. I only  H. III only
G. II only  I. I and II

**13.** Cindy is reflecting quadrilateral *ABCD* over the *y*-axis. Which coordinates represent the image of *D*?

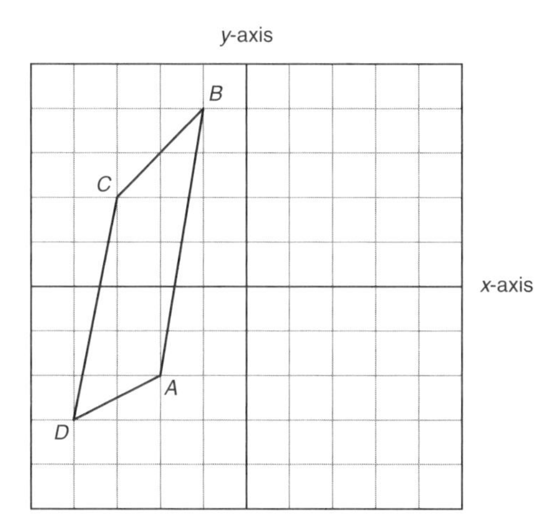


**A.** (–3, 4) **C.** (–4, 3)
**B.** (4, –3) **D.** (–4, –3)

**14.** Lines *AB* and *EF* in the graph will intersect at what coordinates?

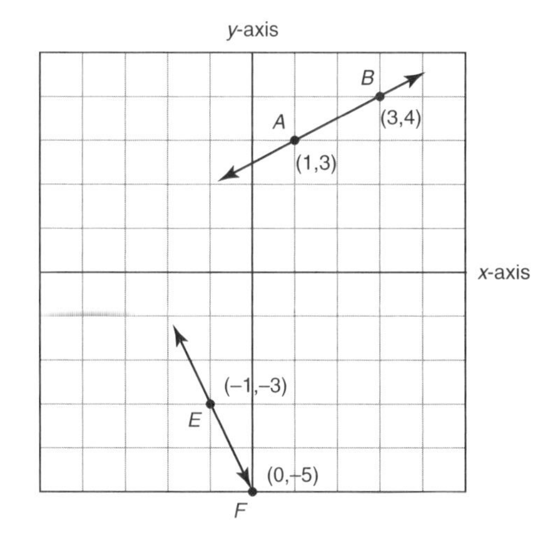


**F.** (–3, –1) **H.** (–1, –3)
**G.** (–3, 1) **I.** (3, –1)

**15.** Which equation can be used to generate this function table?

| $x$ | 2 | 4 | 6 | 8 |
|---|---|---|---|---|
| $y$ | 12 | 18 | 24 | 30 |

**A.** $y = x + 10$ **C.** $y = x + 4$
**B.** $y = 3x + 6$ **D.** $y = 6x$

**16.** Which equation generates the line on the graph?

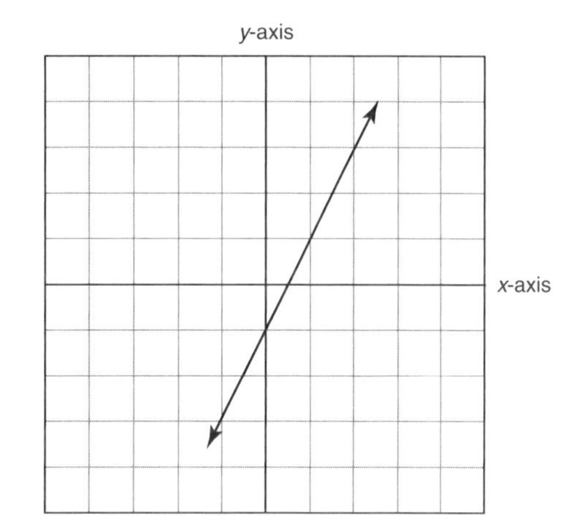


**F.** $y = 2x - 1$ **H.** $y = 2x + 1$
**G.** $y = \frac{1}{2}x - 1$ **I.** $y = \frac{1}{2} + 1$

**17.** Keesha's mother is a salesperson who earns an annual salary of \$12,000 plus a 25% commission on total sales *s*. Which equation can be used to determine her income *i* if she sells \$72,000 in merchandise during a calendar year?

**A.** $i = 72{,}000s + 12{,}000$
**B.** $i = 12{,}000s + 72{,}000$
**C.** $i = 0.25(72{,}000) + 12{,}000$
**D.** $i = s + 12{,}000(0.25)$

**18.** A country buffet restaurant charges \$6.50 for a lunch buffet plate and \$1.50 for a beverage. A party of four eats lunch at this restaurant. Each person orders a buffet and a beverage or a free water. If the party spends a total of \$29 before tax, how many people ordered beverages?

F. 1
G. 2
H. 3
I. 4

**19.** The bar graph shows the population densities of five Florida counties by displaying how many persons inhabit each county per square mile. Which title should be used for the vertical axis?

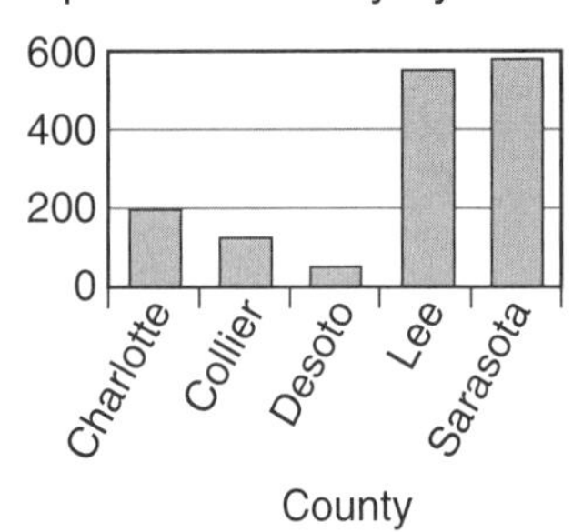


A. Population (thousands)
B. Number of Density
C. People per Square Mile
D. Land Area (square miles)

**20.** The box-and-whiskers plot shows the yearly salaries of 13 employees in a company. Which measures can be obtained from the plot?

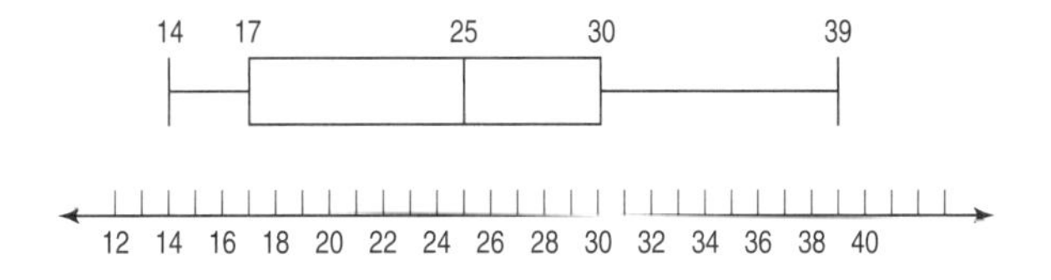


F. mean and range
G. mean and median
H. median and range
I. median and mode

**21.** Mary's purse contains 3 quarters, 2 dimes, 4 nickels, and 6 pennies. If she takes out a coin at random, what are the odds that it is NOT a penny?

A. 3 to 2
B. 2 to 3
C. 3 to 6
D. 6 to 4

22. The circle graph represents the composition of methane gas. The chemical composition of methane gas is $CH_3$, meaning one carbon atom for every three hydrogen atoms. Kyle drew the circle graph to represent the relationship between the carbon atoms and hydrogen atoms in the gas. Which is an accurate evaluation of this graph?

Carbon Atoms vs. Hydrogen Atoms in Methane Gas

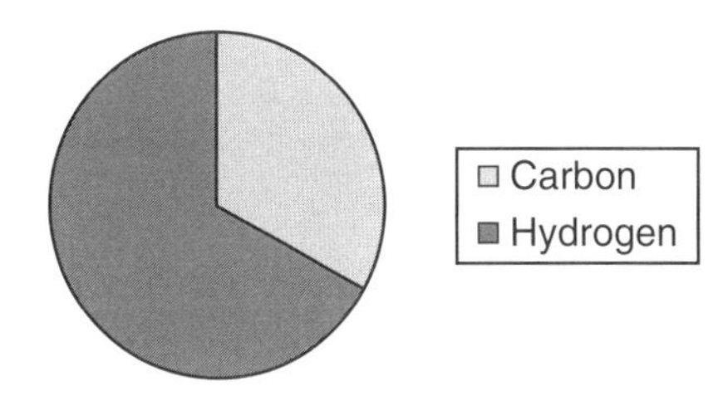


F. The graph is accurate.

G. The graph should have more than two sections.

H. The graph is incorrect because $\frac{1}{4}$ of the graph should represent carbon and $\frac{3}{4}$ should represent hydrogen.

I. The graph is incorrect because there should be more carbon than hydrogen.

23. A science student in a chemistry class is weighing a sample of material. The mass of the substance is $6.5 \times 10^{-2}$. What is the decimal equivalent of this number?

24. Simplify $3^5 + 5^3$.

25. What is the value of the expression $6 \times 8 - 9 \div 3$?

26. The accompanying table shows the original purchase price and discount for each of four stereos. What is the least amount, in dollars, that a customer could pay for one of these stereos before tax?

Stereos

| Brand | Price | Discount |
|---|---|---|
| A | \$220 | Save \$130 |
| B | \$178 | 50% off |
| C | \$199 | Save \$100 |
| D | \$139 | 40% off |

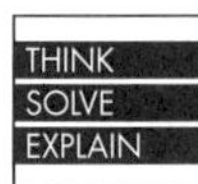


27. A new football stadium has 4 sections. Each section has 21 rows with 98 seats in each row. What is the ESTIMATED total number of seats in the stadium? Explain the techniques that you used to determine your answer.

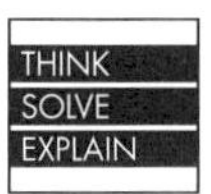


28. A can of soup has the dimensions shown.

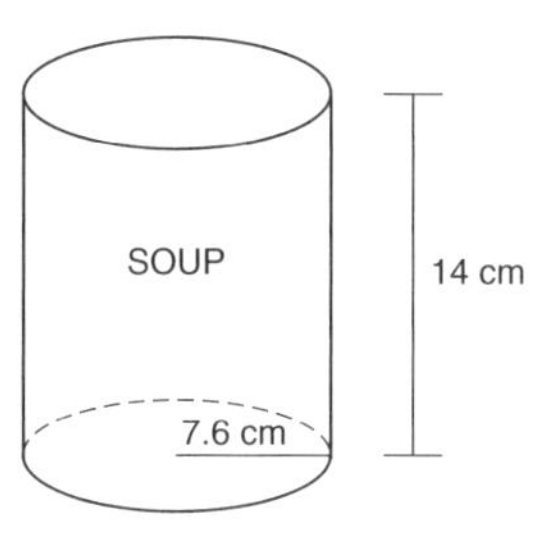


Part A. Use the dimensions of the container to determine its capacity. Explain your reasoning, or show how you arrived at your answer.

Part B. If 1 liter = 1000 cubic centimeters, how many liters of soup will the container hold?

29. A toy company markets a scale-model Chevy S-10 truck with a truck bed capacity of 64 cubic inches. The actual dimensions of the truck are 15 times bigger than this model. What is the capacity in cubic feet of the truck bed on the actual truck?

| | | | | |
|---|---|---|---|---|
| | / | / | / | |
| . | . | . | . | . |
| 0 | 0 | 0 | 0 | 0 |
| 1 | 1 | 1 | 1 | 1 |
| 2 | 2 | 2 | 2 | 2 |
| 3 | 3 | 3 | 3 | 3 |
| 4 | 4 | 4 | 4 | 4 |
| 5 | 5 | 5 | 5 | 5 |
| 6 | 6 | 6 | 6 | 6 |
| 7 | 7 | 7 | 7 | 7 |
| 8 | 8 | 8 | 8 | 8 |
| 9 | 9 | 9 | 9 | 9 |

30. In the regular pentagon shown, find the measure, in degrees, of $\angle 1$.

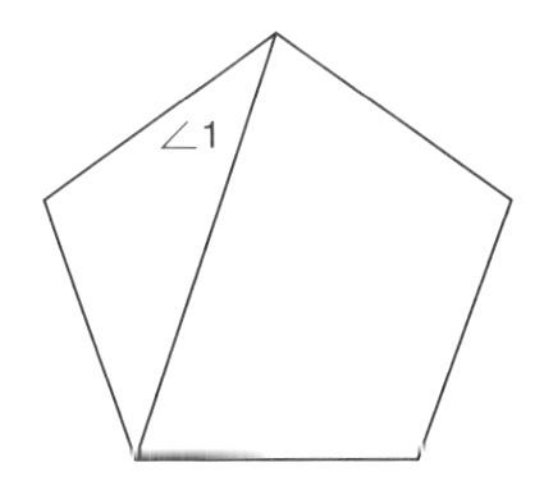


| | | | | |
|---|---|---|---|---|
| | / | / | / | |
| . | . | . | . | . |
| 0 | 0 | 0 | 0 | 0 |
| 1 | 1 | 1 | 1 | 1 |
| 2 | 2 | 2 | 2 | 2 |
| 3 | 3 | 3 | 3 | 3 |
| 4 | 4 | 4 | 4 | 4 |
| 5 | 5 | 5 | 5 | 5 |
| 6 | 6 | 6 | 6 | 6 |
| 7 | 7 | 7 | 7 | 7 |
| 8 | 8 | 8 | 8 | 8 |
| 9 | 9 | 9 | 9 | 9 |

31. A dollhouse refrigerator was made using a scale of 1:12. It has a capacity of 15 cubic inches. By what number would you multiply to find the capacity, in cubic inches, of the actual refrigerator?

A. 12
B. 24
C. 144
D. 1728

32. A campus map shows that the distance from the east door of Todd's building to the west door is approximately 3 inches. If the map has a scale of 2 inches = 50 feet, what is the actual distance, in feet, between the doors?

| | | | | |
|---|---|---|---|---|
| | / | / | / | |
| . | . | . | . | . |
| 0 | 0 | 0 | 0 | 0 |
| 1 | 1 | 1 | 1 | 1 |
| 2 | 2 | 2 | 2 | 2 |
| 3 | 3 | 3 | 3 | 3 |
| 4 | 4 | 4 | 4 | 4 |
| 5 | 5 | 5 | 5 | 5 |
| 6 | 6 | 6 | 6 | 6 |
| 7 | 7 | 7 | 7 | 7 |
| 8 | 8 | 8 | 8 | 8 |
| 9 | 9 | 9 | 9 | 9 |

33. Wendy's father spends 2 hours a day 5 days a week commuting back and forth to work. In 4 weeks, how many minutes does he spend commuting?

| | | | | |
|---|---|---|---|---|
| | / | / | / | |
| . | . | . | . | . |
| 0 | 0 | 0 | 0 | 0 |
| 1 | 1 | 1 | 1 | 1 |
| 2 | 2 | 2 | 2 | 2 |
| 3 | 3 | 3 | 3 | 3 |
| 4 | 4 | 4 | 4 | 4 |
| 5 | 5 | 5 | 5 | 5 |
| 6 | 6 | 6 | 6 | 6 |
| 7 | 7 | 7 | 7 | 7 |
| 8 | 8 | 8 | 8 | 8 |
| 9 | 9 | 9 | 9 | 9 |

34. What is the measure, in degrees, of angle $a$?

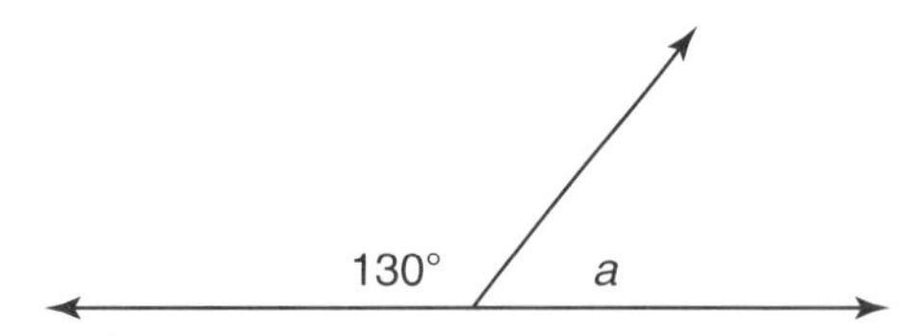


35. In the figure, trapezoid $ABCD$ is dilated by what scale factor?

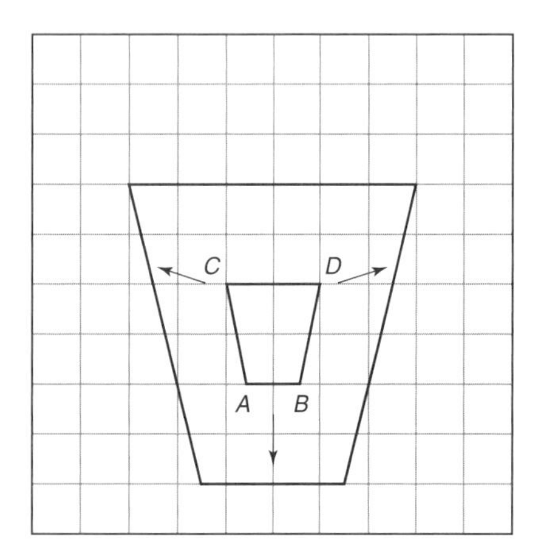


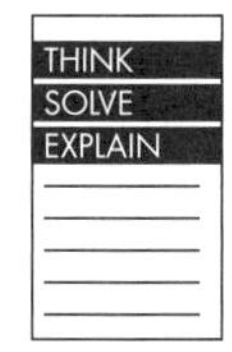


36. On the grid, perform the following transformations:

Part A. Translate thc triangle $DEF$ 4 units to the left.

Part B. Rotate the figure 90 degrees clockwise about point (–2, –2).

Part C. Reflect the figure over the $y$-axis, and relabel as $D'E'F'$. What is the relationship between $\overline{DE}$ and $\overline{D'E'}$?

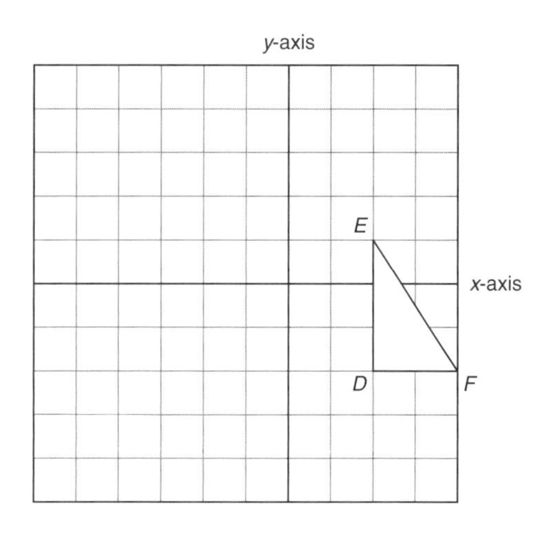


______________________________________

______________________________________

______________________________________

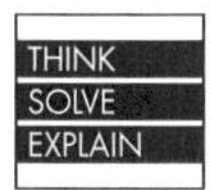


**37.** Are the two figures shown on the grid similar? Why or why not? Explain in words, and use mathematical terms in your answer.

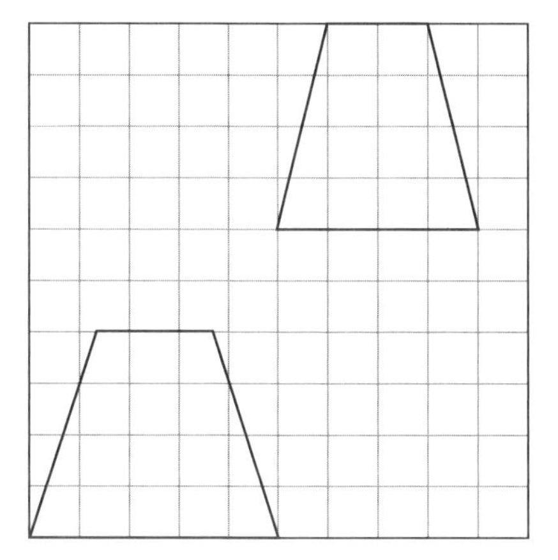

________________________________

________________________________

________________________________

THINK
SOLVE
EXPLAIN

**38.** When a label is wrapped securely around a can of catfood, the label fits perfectly with no overlaps. What expression represents the area of the label? Explain in words how to write the expression.

________________________________

________________________________

________________________________

**39.** Sophia is heating wax to make candles. In 3 minutes, the wax has heated to 100°F. After 7 minutes, the temperature of the wax is 120°F.

Part A. Plot these coordinates on the grid, and draw the connecting ray.

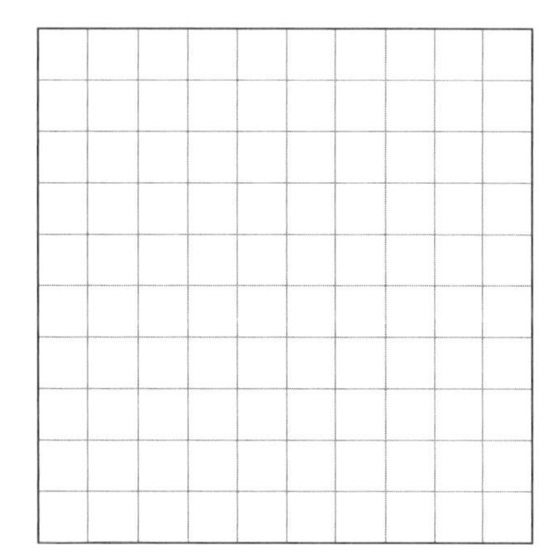

Part B. Assume that the wax continues to heat at a steady rate. Extend your graph to determine the temperature of the wax after 13 minutes.

40. What is the next number in this sequence: 4.75, 5, 5.25, 5.5?

41. A company executive graphs the costs to manufacture different quantities of sewing machine parts. What is the cost, in dollars, to manufacture 30 of these parts?

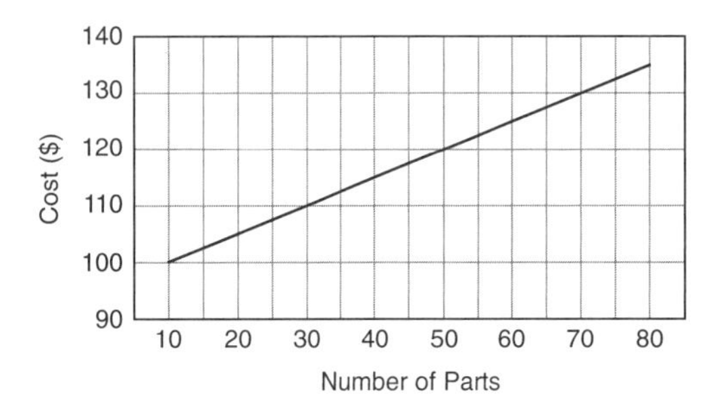


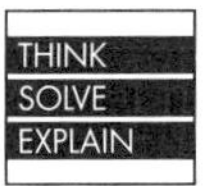


42. Doug is training for a track meet. He makes a table, as shown, of the number of miles he plans to run by the end of each week of training.

| Week | 1 | 2 | 3 | 4 |
|---|---|---|---|---|
| Miles | 5 | 7 | 9 | 11 |

Part A. Write an equation to represent the number of miles $m$ Doug runs in terms of weeks $w$, and use your equation to predict the number of miles he will run during week 10.

______________________________

______________________________

______________________________

Part B. On the grid, draw a graph of the equation you wrote in Part A, and extend it to 10 weeks.

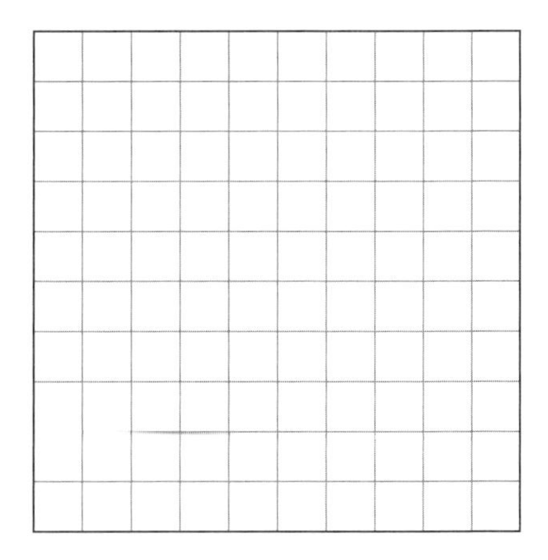

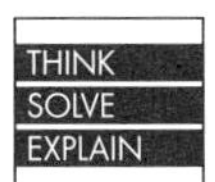


**43.** The graph shows the charges at an airport parking lot.

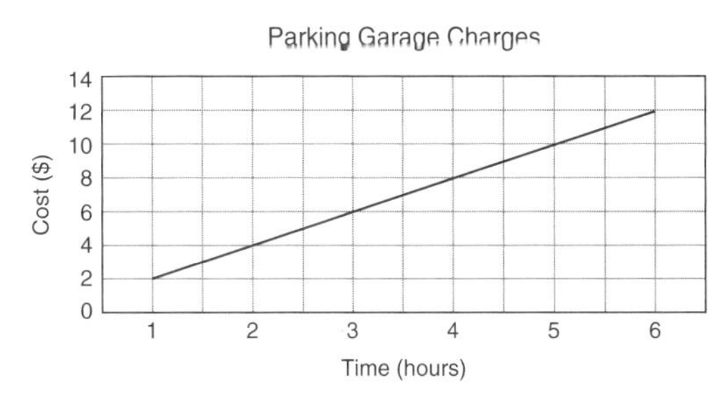


Part A. Write an equation to model this graph. Use *h* for hour.

Part B. Use your equation to calculate the charge for parking 36 hours. Show your work, or explain your steps.

______________________________

______________________________

______________________________

**44.** When Sam drives from his house in Frostproof to Joe's house, the distance is 12 miles. The distance from Joe's house to Leigh Ann's house is 3 times this distance less 5 miles. How many miles is it from Sam's house to Leigh Ann's house?

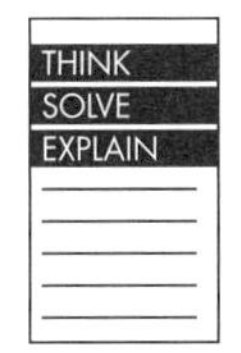


**45.** The Mulray Development Company plans to build a large shopping center in either Lassiter or Stuckey. The company wants to build where the greatest population increase has occurred.

The developers found the population information shown in the table for the two towns over 30 years.

**Population Data (in thousands)**

| Year | Lassiter | Stuckey |
|---|---|---|
| 1970 | 125 | 120 |
| 1980 | 135 | 127 |
| 1990 | 145 | 150 |
| 2000 | 150 | 157 |

Part A. Use the information in the table to construct on the grid a double-line graph to show the difference in growth between the two cities.

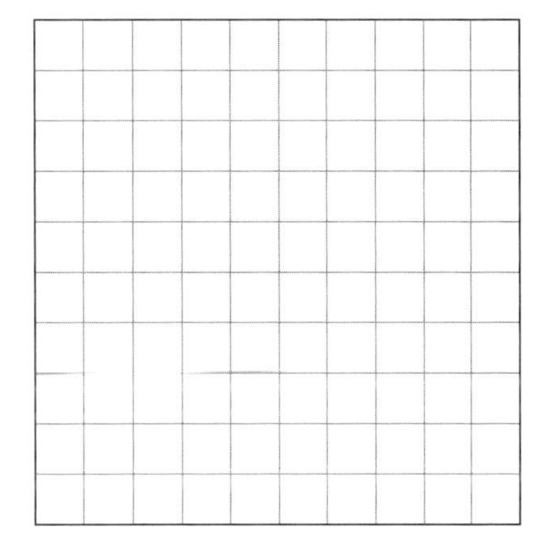

Part B. Compare the information provided by the table and the graph. Which city should the company select to build the shopping center?

Part C. Explain in words how you made your decision in Part B, using data from both the table and your graph.

46. The box-and-whiskers plot shows the data for the reading tests in Mr. Moore's classroom. What is the range of scores?

50 65 74 80 100

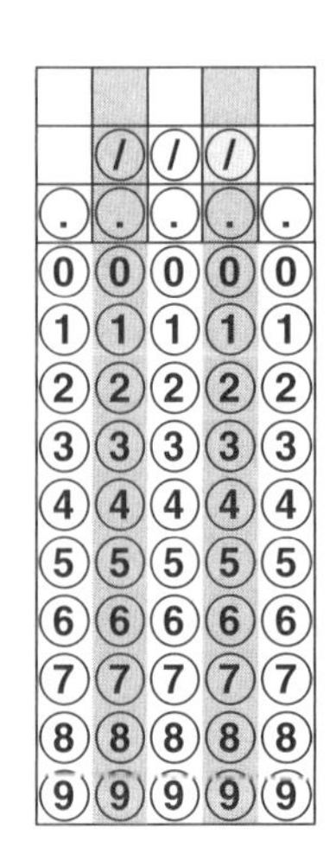

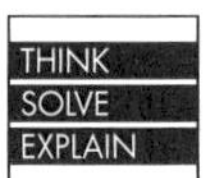


47. A round target board is divided into four equal sections. Shawn stands 20 feet away from the board and throws 20 darts.

Part A. Theoretically, how many darts should end up in each separate section of the board?

Part B. When all the darts have been thrown, there are 4 darts in section 1, 6 darts in section 2, 5 darts in section 3, and 5 darts in section 4. How does this result compare with the theoretical probability you found in Part A?

48. A new popcorn machine purchased by the school for special events burns about 2500 kernels of corn for every 20,000 that are popped. What is the probability that a kernel of popped corn selected at random will NOT be burnt? Express your answer as a fraction.

49. The heights, in inches, of a sample of ten athletes are listed below. The range and accurate measures of central tendency are shown below the heights. Which is the **best** conclusion based on the data set?

    Heights: 65, 68, 70, 72, 68, 74, 73, 69, 71, 70

    Range: 9; mode: 68 and 70; mean: 70; median: 70

    F. All athletes are between 65 and 74 inches tall.
    G. Most athletes are 70 inches tall.
    H. On average, athletes are about 70 inches tall.
    J. No athletes are shorter than 68 inches.

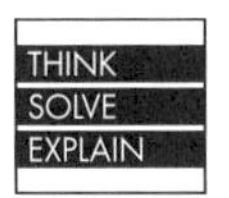


50. Shannon's office held a drawing for baseball tickets. Forty people entered the drawing. In order to win, an employee had to be present at the drawing. Shannon's supervisor drew five names before Shannon's, and none of the people whose names were drawn was present. The names of those not present were eliminated.

    Part A. Determine the probability of Shannon's winning on the sixth draw.

    Part B. Justify your answer by explaining in mathematical terms or showing your work.

    ______________________________

    ______________________________

    ______________________________

## ANSWERS TO PRACTICE TEST ONE

1. **D.** Change each fraction to a decimal: $\frac{2}{3} = 0.67$; $\frac{1}{4} = 0.25$; $\frac{1}{2} = 0.5$; $\frac{3}{4} = 0.75$. Compare the decimals.

2. **G.** $8^4 = 8 \times 8 \times 8 \times 8 = 4096$.

3. **C.** $[(abc)^3]^4 = (a^3b^3c^3) \times (a^3b^3c^3) \times (a^3b^3c^3) \times (a^3b^3c^3) = a^{12}b^{12}c^{12}$.

4. **F.** The circles cross-cancel.

5. **C.** Division and multiplication are performed first, working from left to right: $3 + 4.5 \times 6 - 7 = 3 + 27 - 7$. Next, addition and subtraction are performed, working from left to right: $3 + 27 - 7 = 30 - 7 = 23$.

6. **G.** Use an equation: Jack's age, 15, equals $\frac{1}{3}$ × his mother's age $m + 3$.

$15 = \frac{1}{3}m + 3$

$15 - 3 = \frac{1}{3}m + 3 - 3$

$12 = \frac{1}{3}m$

$36 = m$

7. **B.** Use the formula $A = lw$ for the area of the rectangular driveway. When both the length $\ell$ and the width $w$ are doubled, the new formula is $A = 2\ell \times 2w = 4\ell w$. The new area is 4 times as large.

8. **H.** Change kilometers to meters: 8.4 km = 8400 m. Divide by 13: $8400 \div 13 = 646$.

9. **D.** $17{,}000{,}000 \div 559 = 30{,}411$.

10. **H.** The triangles of the square pyramid net are isosceles.

11. **D.** The angles in a triangle add to 360°. Subtract the given meaures: $180 - (120 + 40) = 20°$

12. **I.** A 180° turn followed by a slide.

13. **B.** The image of $D$ is located at (4, –3), as seen in the diagram.

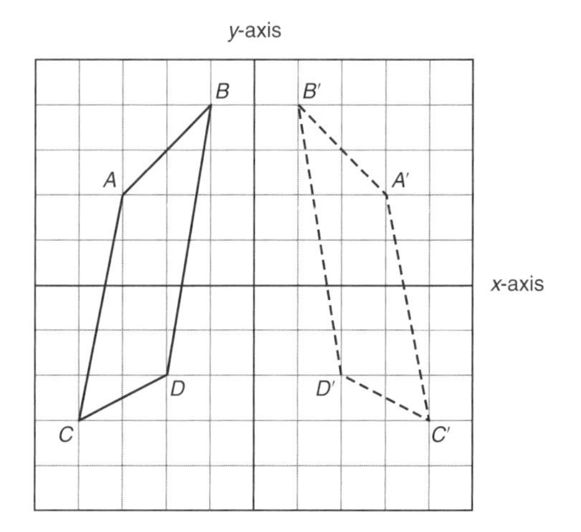

14. **G.**

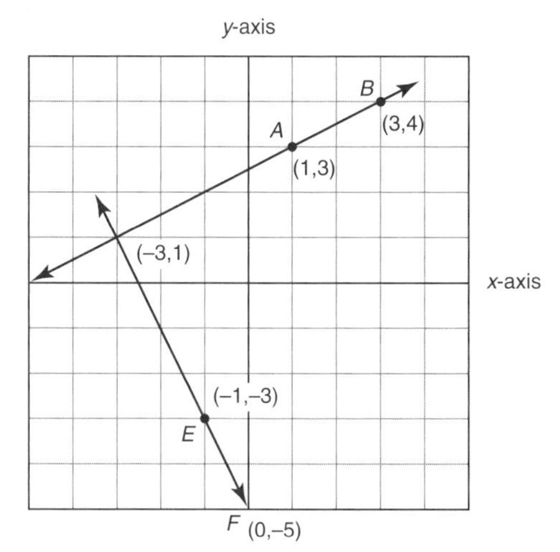


15. **B.** Select an equation, and test it on more than one ordered pair. While $y = x + 10$ will work for the first ordered pair, it will not work for the others. Continue to test. Guess and check is a good strategy for this type of problem.

16. **F.** The $y$-intercept is −1, and the slope of the line is 2.

17. **C.**

18. **G.** Write the equation:
29 = 4(6.5) + beverages
Then 29 = 26 + beverages.
Subtract: 29 – 26 = beverages.
Since $3 was spent for beverages, 2 people must have ordered 2 beverages.

19. **C.** Read carefully: ". . . how many persons inhabit cach county per square mile."

20. **H.** Box-and-whiskers plots show median and range.

21. **A.** Mary's purse contains a total of 15 coins. Of these 9 are not pennies and 6 are pennies. The ratio 9:6 (in lowest terms 3:2) shows the odds that the coin *is not* a penny.

22. **H.** The correct graph should reflect the fact that there are 4 atoms: 1 is carbon and 3 are hydrogen, as shown below.

Carbon Atoms vs. Hydrogen Atoms in Methane Gas

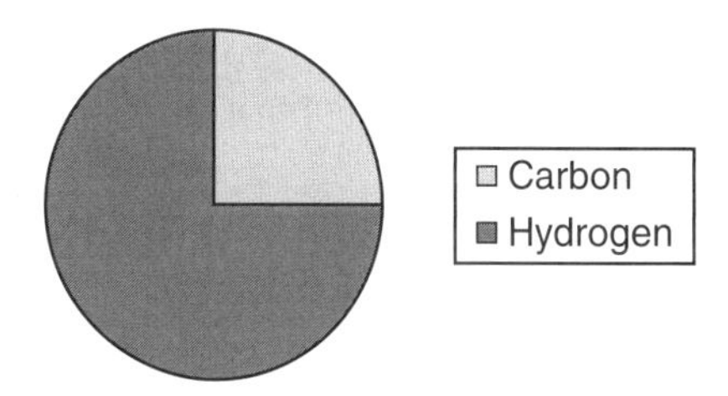


23. **0.065.** Multiplying by $10^{-2}$ moves the decimal point two places to the left.

24. **368.** The numbers have different bases. Evaluate separately, then add: $3^5 = 243$, $5^3 = 125$; $243 + 125 = 368$.

25. **45.** First do all multiplications and divisions, then subtract.
$6 \times 8 - 9 \div 3 = 48 - 9 \div 3 =$
$48 - 3 = 45$.

26. **83.40.** Use the percentage key on your calculator: $139 – 40%.

| Brand | Price | Discount | Amount Paid |
|---|---|---|---|
| A | $220 | Save $130 | $220 – 130 = $90.00 |
| B | $178 | 50% off | $178 ÷ 2 = $89.00 |
| C | $199 | Save $100 | $199 – 100 = $99.00 |
| D | $139 | 40% off | $139 – 40% = $83.40 |

27. **8000.** To estimate, round before beginning calculations. Round 21 to 20 and 98 to 100. Multiply: $4 \times 20 \times 100 = 8000$.

28. Part A. **2539.1 cu cm.** To find capacity, use the volume formula for a cylinder: $V = \pi r^2 h$.

    $V = \pi r^2 h$
    $V = 3.14 \times r \times r \times h$. The radius is 7.6 cm, and the height is 14 cm.
    $V = 3.14 \times 7.6 \times 7.6 \times 14 =$ 2539.1 cu cm.

    Part B. **2.539 L.** Since 1 L = 1000 cu cm, $2539.1 \div 1000 = 2.539$ liters.

29. **125.** A truck bed is three-dimensional: length, width, and height. To find the capacity of the actual truck, multiply the capacity of the scale-model truck (64 cubic inches) by 15 three times: $64 \times 15 \times 15 \times 15 = 216{,}000$ cubic inches. Next, convert cubic inches to cubic feet by dividing by 12 three times (once for each dimension): $216{,}000 \div 12 \div 12 \div 12 = 125$ cubic feet.

30. **38.** In a regular pentagon, each angle measures 108°. This is found by using the formula for interior angles of a polygon shown on the FCAT Reference Sheet, where $n$ = number of sides in the polygon:

    $$\frac{180(n-2)}{n} = \frac{180(5-2)}{5} = \frac{180(3)}{5} = \frac{540}{5} = 108.$$

    To find the measure of $\angle 1$, subtract: $180° - 108° = 72°$. Divide 72° by 2.

31. **D.** Each dimension of the dollhouse refrigerator (length, width, and height) must be multiplied by 12: $12 \times 12 \times 12 = 1728$.

32. **75 ft.** Use a proportion to solve this problem: $\frac{3 \text{ in}}{\text{actual distance}} = \frac{2 \text{ in}}{50 \text{ ft}}$. Multiply $3 \times 50$, and divide by 2: $150 \div 2 = 75$.

33. **2400.** The answer must be in minutes; therefore, convert 2 hr to minutes by multiplying by 60. Then, $120 \times 5 \times 4 = 2400$.

34. **50.** The two angles in the diagram are supplementary and must add to 180°: $180° - 130° = 50°$.

35. **3.** The smaller figure has a height of 2 and a width of 2. The larger figure has a height of 6 and a width of 6. Trapezoid *ABCD*'s dimensions are dilated by a factor of 3.

36.

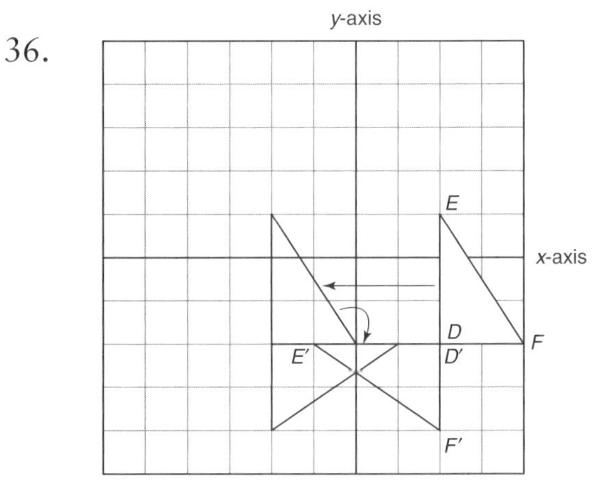


$\overline{DE}$ and $\overline{D'E'}$ are **perpendicular**. See transformations *A*, *B*, and *C*.

37. **No.** The two figures shown are not similar because similar figures are proportional. The bottom figure has a base of 5 and a height of 4. The top figure has a base of 4 and a height of 4. The base and height were not reduced proportionally: $\frac{5}{4} \neq \frac{4}{4}$.

38. **$\pi dh$.** The area of the label is found by multiplying length by width. The length of the label is actually the circumference $\pi d$ of the can, while the width of the label is actually the height $h$ of the can.

39. Part A.

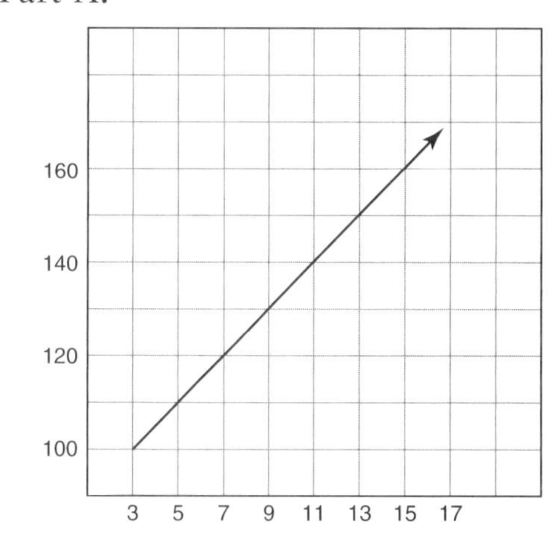


Part B. According to the graph, after 13 minutes the temperature of the wax will be **150°**.

40. **5.75.** The sequence is increasing by 0.25 each time.

41. **110.**

42. Part A. **$m = 2w + 3$, 23.** The number of miles Doug runs increases by 2 each week; therefore, multiply weeks $w$ by 2:
$m = 2(10) + 3 = 20 + 3 = 23$.

Part B.

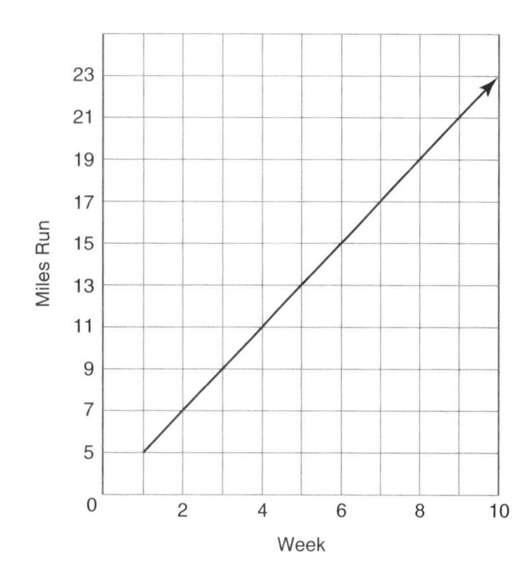


43. Part A. **$C = 2h$.**

Part B. **$72.** Charge $C = 2(36) = \$72$.

44. **31.** The distance from Joe's house to Leigh Ann's house is 3 times 12 minus 5: $3(12) - 5 = 31$.

45. Part A.

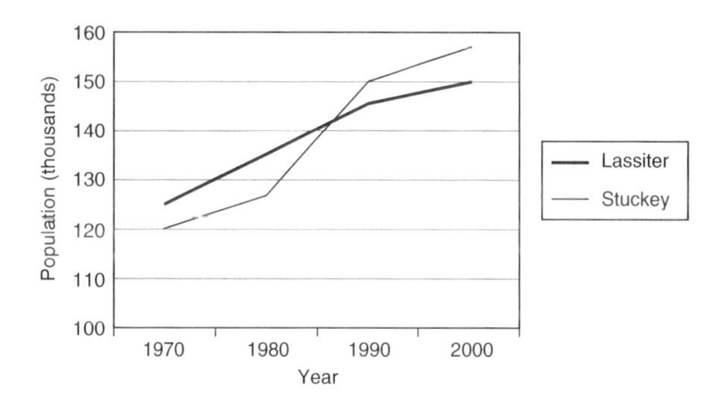

Part B. **Stuckey**

Part C. Although Stuckey started out lower than Lassiter, its population grew and passed the population of Lassiter in 1985. Thereafter, it continued to grow at a rate slightly faster than that of Lassiter.

46. **50.** The range is the difference between the highest and lowest scores: $100 - 50 = 50$.

47. Part A. **5.** Theoretically, 5 darts should end up in each section of the board.

Part B. The theoretical and experimental probabilities are actually pretty close. Section 1 has 4 darts instead of 5, and Section 2 has 6 darts instead of 5. Sections 2 and 3 both have 5, the theoretical probability.

48. $\frac{7}{8}$. If 2500 kernals are burnt, then 17,500 will not be burnt. The probability is: $\frac{17{,}500}{20{,}000} = \frac{175}{200} = \frac{7}{8}$.

49. **H.** Although A is a tempting answer, it is important to remember that 10 athletes represent only a small sample of all athletes.

50. Part A. $\mathbf{\frac{1}{35}}$.

Part B. Once there have been 5 drawings, only 35 names are left. Therefore, Shannon has a 1 in 35 chance of winning on the sixth draw.

# Practice Test 2

1. Which number has the greatest value?

   A. $3^3$ C. $26\frac{7}{8}$

   B. $\sqrt{730}$ D. 27%

2. In lowest terms, what fraction of the square shown is shaded?

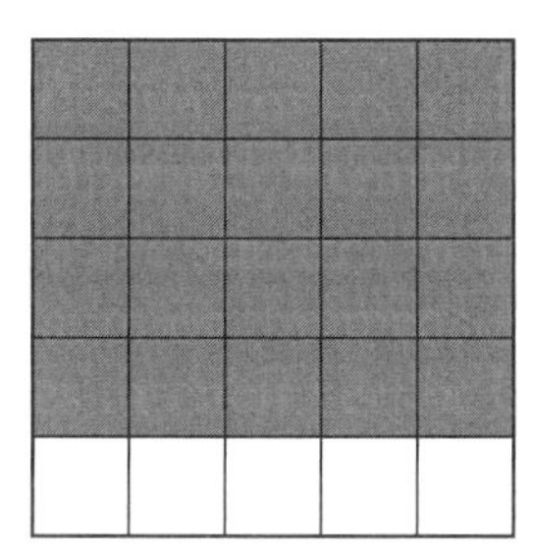

3. The world population is projected to grow at about 70 million people per year. Which is NOT another way to express the number 70 million?

   F. 70,000,000
   G. $7.0 \times 10^7$
   H. 70,000,000,000
   I. $70 \times 1{,}000{,}000$

4. Find $\frac{10^9}{10^5}$.

   A. $1^{14}$ C. $10^{14}$
   B. $1^4$ D. $10^4$

5. Find $(4.5 \times 10^{-7}) \cdot (3.0 \times 10^8)$.

6. The additive inverse of 16 is which kind of number?

   F. a negative integer such that the product of 16 and the number is 1
   G. a number such that the sum of 16 and the number is 16
   H. a positive integer such that the product of 16 and the number is 1
   I. a negative integer such that the sum of 16 and the number is 0

7. Which operation should be performed **last** to simplify the expression $87 - 3(6 \times 2)(20 - 18)$?

   A. $87 - 72$
   B. $6 \times 2$
   C. $20 - 18$
   D. $87 - 3$

8. The expression $\frac{7 \times 2(24 \div 3)}{20 + 14 - 6}$ has what integer value?

|   |   |   |   |   |
|---|---|---|---|---|
|   | / | / | / |   |
| . | . | . | . | . |
| 0 | 0 | 0 | 0 | 0 |
| 1 | 1 | 1 | 1 | 1 |
| 2 | 2 | 2 | 2 | 2 |
| 3 | 3 | 3 | 3 | 3 |
| 4 | 4 | 4 | 4 | 4 |
| 5 | 5 | 5 | 5 | 5 |
| 6 | 6 | 6 | 6 | 6 |
| 7 | 7 | 7 | 7 | 7 |
| 8 | 8 | 8 | 8 | 8 |
| 9 | 9 | 9 | 9 | 9 |

9. A lemon shark found off Manasota Key had a length of $14\frac{7}{8}$ feet. Another lemon shark found further up the coast near Tampa Bay measured $9\frac{1}{2}$ feet. What is the difference in length between the two sharks?

   F. 5 ft $\frac{5}{8}$ in
   G. 5 ft $\frac{3}{8}$ in
   H. 7 ft
   I. 7 ft $\frac{3}{8}$ in

10. A company manufactures 10,000 specialty tires in 4 weeks. If 5% are defective and cannot be sold, how many tires can the company sell in 1 week?

|   |   |   |   |   |
|---|---|---|---|---|
|   | / | / | / |   |
| . | . | . | . | . |
| 0 | 0 | 0 | 0 | 0 |
| 1 | 1 | 1 | 1 | 1 |
| 2 | 2 | 2 | 2 | 2 |
| 3 | 3 | 3 | 3 | 3 |
| 4 | 4 | 4 | 4 | 4 |
| 5 | 5 | 5 | 5 | 5 |
| 6 | 6 | 6 | 6 | 6 |
| 7 | 7 | 7 | 7 | 7 |
| 8 | 8 | 8 | 8 | 8 |
| 9 | 9 | 9 | 9 | 9 |

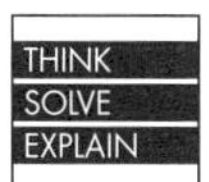


**11.** The Everglades provides refuge for over 68 endangered species within its 26,000-square-mile area. How can a wildlife conservationist ESTIMATE how many square miles there are per endangered species? Show your work.

**12.** One cubic inch of tile from the space shuttle weighs approximately 0.05 ounce. Ms. Reynolds brings a perfect cube of the shuttle tile with a side length of 4 inches to show her science class. What is the weight, in pounds, of this cube of shuttle tile?

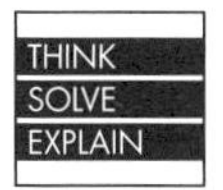


**13.** The post office uses boxes measuring 11 inches by 8.5 inches by 5.5 inches to send Priority Mail. Jay wants to send tangerines to his grandchildren in Alaska.

Part A. If each tangerine occupies about 20 cubic inches, approximately how many tangerines will fit in a box?

Part B. If Jay's picking basket holds 1 cubic foot of tangerines, will a Priority Mail box hold them all? Show your work, or explain how you arrived at your answer.

**14.** An expert typist can type 90 words per minute. How many words can he type in three quarters of an hour?

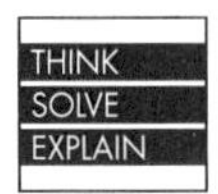


**15.** A cheetah can run about 70 miles per hour for short distances.

Part A. If the cheetah chases its prey for $\frac{1}{2}$ minute, how many miles will the cheetah have traveled? Show your work, or explain how you arrived at your answer.

Part B: Use your answer from Part A to ESTIMATE how many feet the cheetah ran. Show your work, or explain how you arrived at your answer.

**16.** Motel Marine is putting in a new patio for visitors. Its dimenions will be 12 meters by 5 meters. The old patio had half these dimensions. How much more space will visitors have when the new patio is completed?

A. They will have double the area.
B. They will have 4 times the area.
C. They will have 6 times the area.
D. They will have 12 times the area.

**17.** An architect designs a room with dimensions of 10 feet by 12 feet. The owner prefers a more open floor plan and has the architect redesign the room with triple the dimensions. What is the *perimeter*, in feet, of the larger room?

**18.** An architect designs a building and drafts the scale model shown. If the actual building will measure 180 feet wide by 60 feet tall, what does 1 inch represent on the architect's scale?

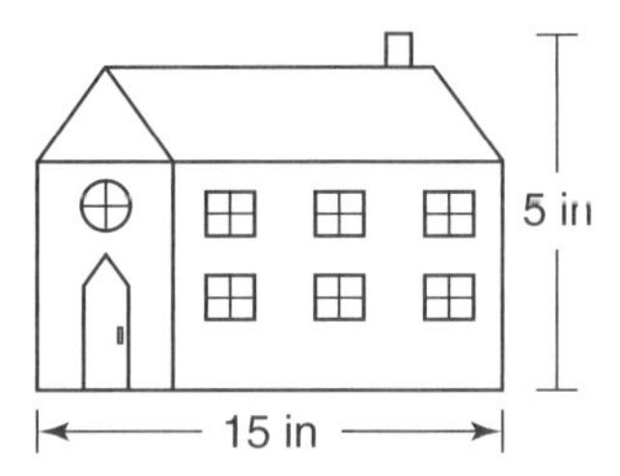


F. 1 ft
G. 8 ft
H. 12 ft
I. 24 ft

**19.** The scale on a map shows $\frac{1}{4}$ inch = 5 miles. If two places are $4\frac{1}{2}$ inches apart on the map, what is the actual distance, in miles, between them?

**20.** For her living room, Lannie purchased a new rug that measured 4.8 meters in length. What was the length of the rug in millimeters?

A. 48
B. 480
C. 4800
D. 48,000

**21.** Sam works at a boat shop for 6 hours a day, 5 days a week. How many **minutes** does Sam work at the shop each week?

**22.** Sandi cuts an orange in the shape of a hemisphere with a radius of 4 centimeters. What is the measure, in centimeters, of the distance shown in the drawing?

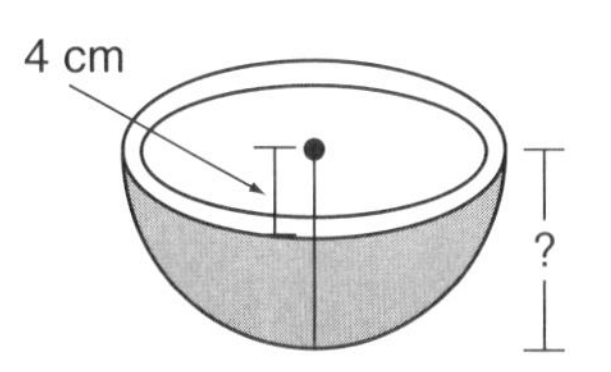


**23.** For the rectangle *ABCD*, which statement explains why $\overline{AC}$ has the **same length** as $\overline{BD}$?

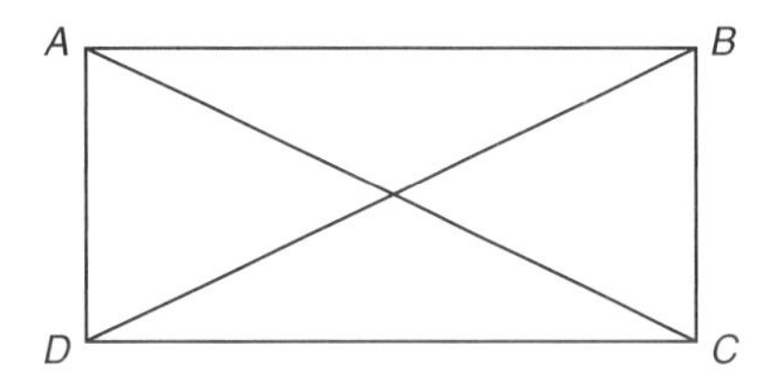


F. Diagonals of a rectangle are congruent.
G. Diagonals of a rectangle bisect each other.
H. Opposite sides of a rectangle are congruent.
I. Diagonals of a parallelogram bisect each other.

24. For quadrilateral $LMPQ$, if $\overleftrightarrow{LM} \parallel \overleftrightarrow{PQ}$, which condition **must** be true?

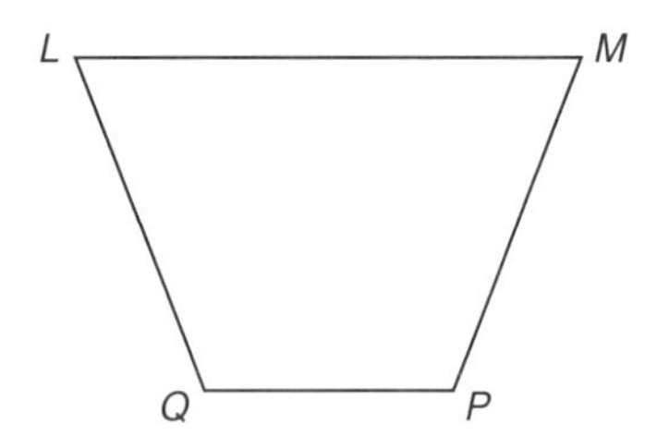


A. $L$, $M$, and $P$ form a triangle.

B. $\overleftrightarrow{LM} \perp \overleftrightarrow{PQ}$

C. $\overleftrightarrow{LP} \perp \overleftrightarrow{PQ}$

D. $\overleftrightarrow{LP} \parallel \overleftrightarrow{MQ}$

25. The isosceles triangles shown are similar. What is the length of the base of the larger triangle?

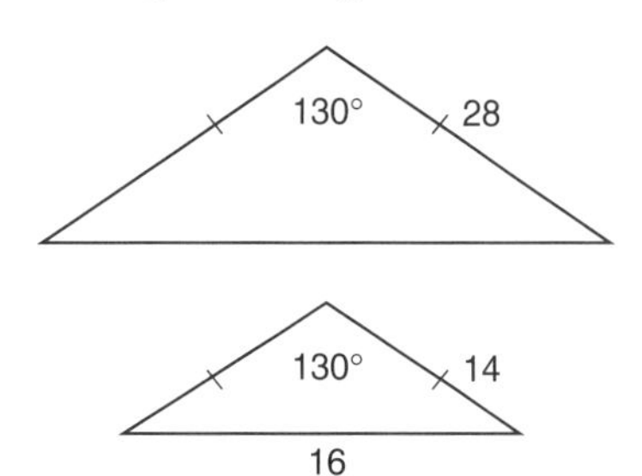


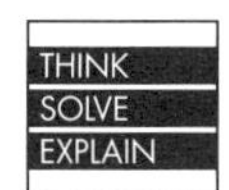


26. Part A. On the grid, plot and label rectangle $ABCD$ with the following conditions:

1. $A$ is at (1, 1).
2. $ABCD$ is not in Quadrant II.
3. The $x$-axis serves as a line of symmetry for the figure.
4. Rectangle $ABCD$ is congruent to rectangle $HIJK$.

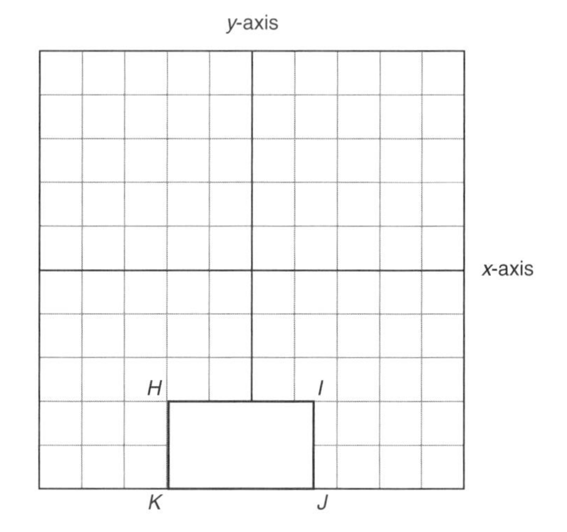


Part B. Explain in words how you determined where to plot rectangle $ABCD$.

______________________________

______________________________

______________________________

27. An Englewood tile shop has a floor sample of tiles in the window as shown. The owner of the shop explains that the design is a tessellation. Why is this sample a tesselation?

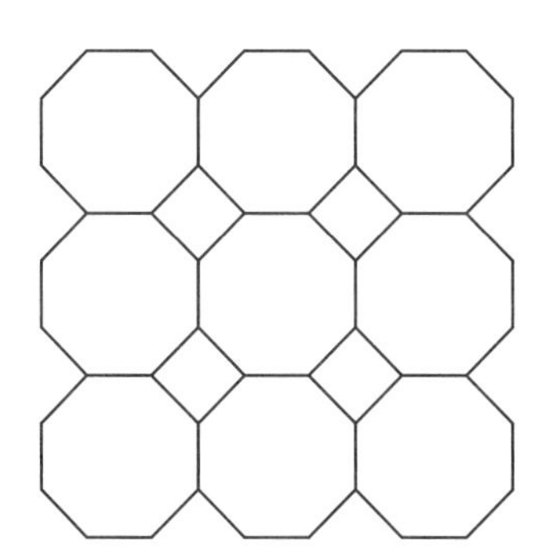

I. The design covers a plane with no overlapping or empty spaces.
II. There are squares in the design.
III. The sum of the measures of the angles at each vertex is 360°.

F. I and III only
G. I only
H. I, II, and III
I. II and III only

28. In which figure is line $b$ a line of symmetry?

A.
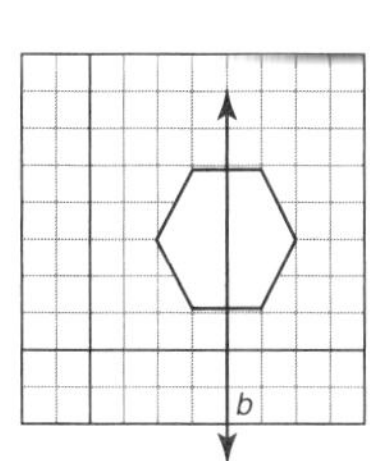


C.
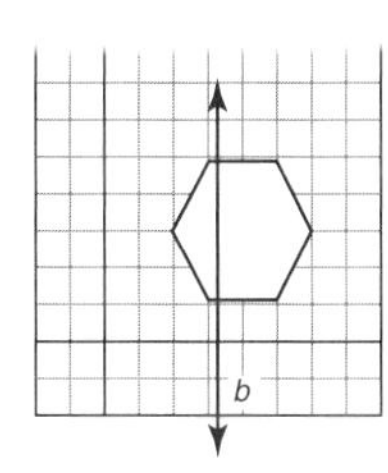


B.
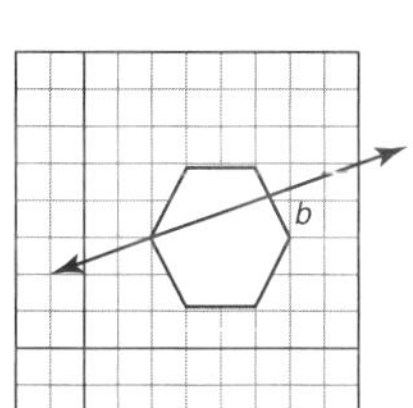


D.
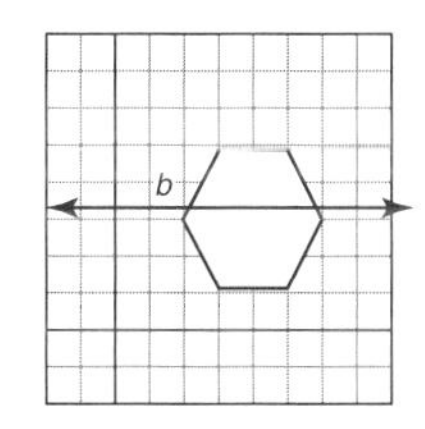


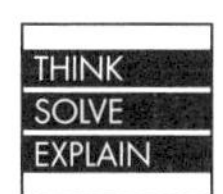


29. Part A: A weather-reporting station is designing a map to show how lightning storms move across the state. For this purpose, the meteorologists use a coordinate grid, as shown. Describe the transformations from lightning $A$ to lightning $B$ and from lightning $B$ to lightning $C$. Use mathematical terms to describe the transformation.

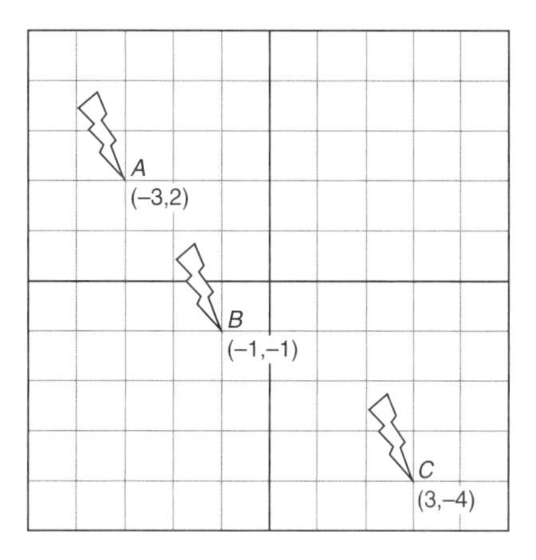


______________________________

______________________________

______________________________

Part B. Use the Pythagorean theorem to estimate the distance, in spaces, between lightning $B$ and lightning $C$.

______________________________

______________________________

______________________________

30. Which line has one intercept and undefined slope?

F. 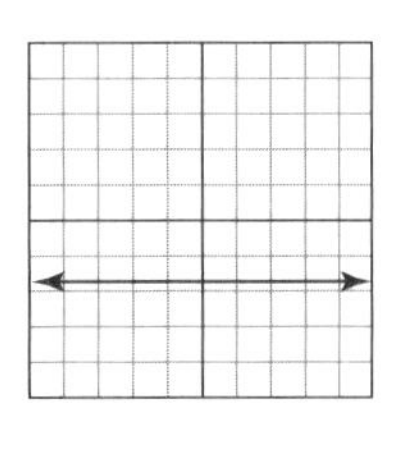

H. 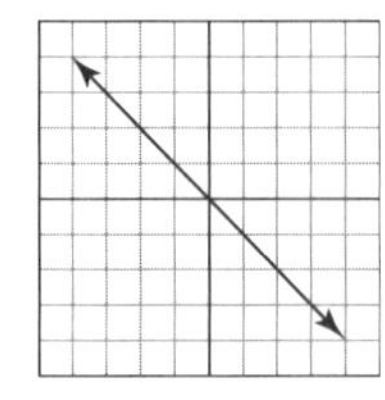

G. 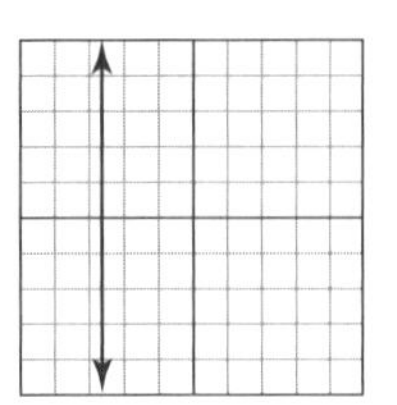

I. 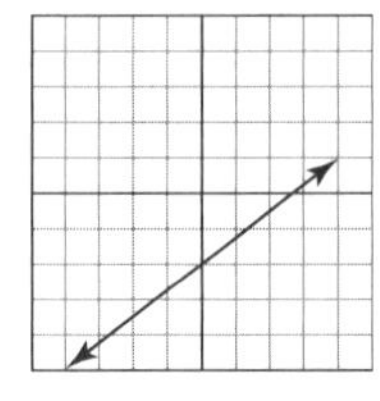

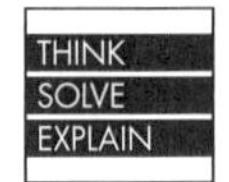


31. Part A. On the grid, plot $A$ at (4, –3) and then construct $\overline{AB}$ with an $x$-intercept of –1. Plot $B$ in a different quadrant from $A$.

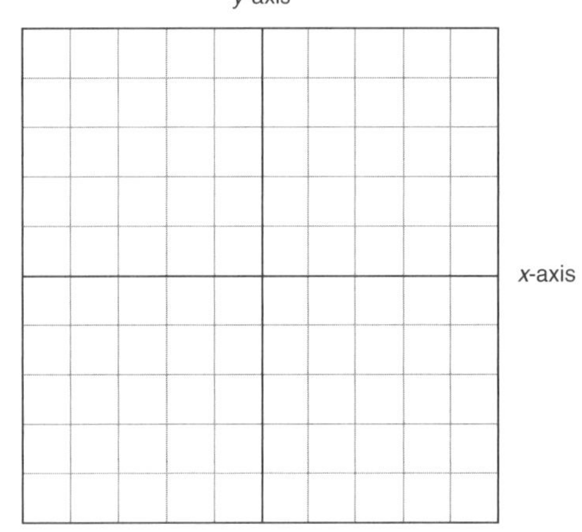


Part B. Name the quadrants that contain $A$ and $B$.

__________________________________

__________________________________

__________________________________

32. Which is NOT a solution of the inequality $y \leq 4x - 6$?

A. (2, 2) C. (4, 12)
B. (5, 4) D. (10, 13)

33. Observe the pattern of triangles and squares in the diagrams shown. If this pattern were continued, how many triangles would there be in a design having 20 squares?

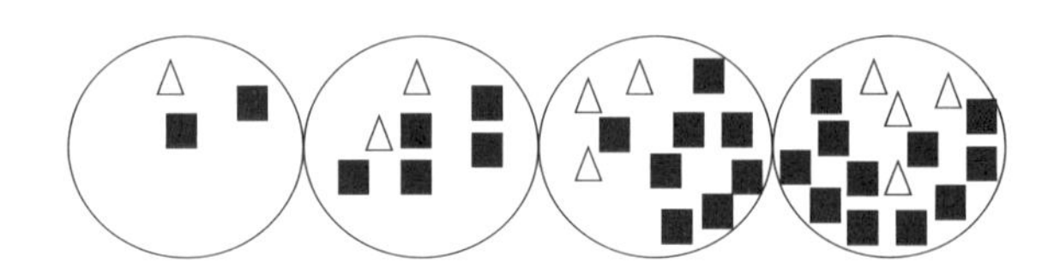

34. Which equation generates the line on the graph?

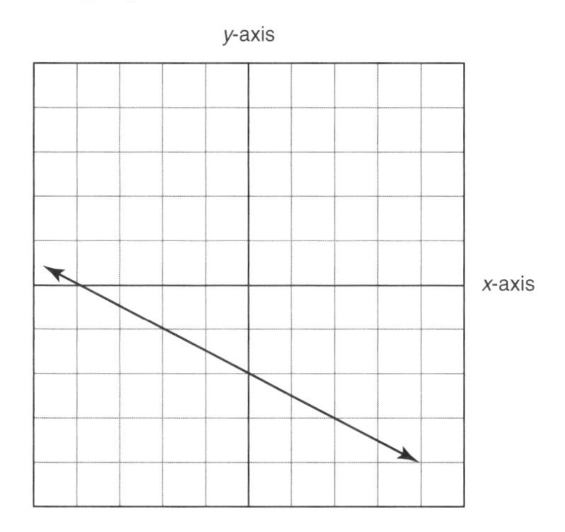


F. $y = \frac{1}{2}x - 2$

G. $y = -\frac{1}{2}x - 2$

H. $y = 2x - 2$

I. $y = -2x - 2$

35. For the equation $y = 15x - 6$, what is the value of $y$ when $x = \frac{1}{2}$?

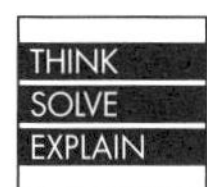


36. Charlie is in the U.S. Navy. He knows that the temperature of the water depends on how deep his submarine dives. The graph shows the temperature of the water as a function of the depth after the first 1000 feet.

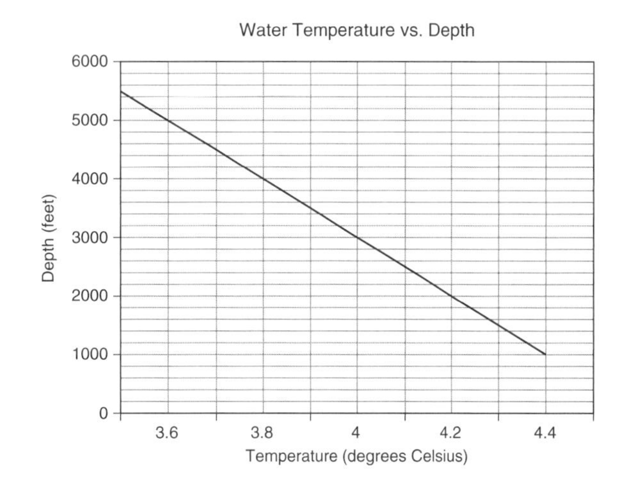


Part A. Find the slope of the line that models the temperature $t$ that Charlie encounters underwater for depth $d$.

Part B. Explain how you wrote your slope, and state its meaning.

________________________________

________________________________

________________________________

37. Which equation can be read aloud as "The difference between half a number and 7 equals the sum of 6 and another number?"

 A. $2n - 7 = 6 + b$

 B. $\frac{1}{2}n - 7 = 6 + b$

 C. $6n - 6 = \frac{1}{2} + b$

 D. $\frac{1}{2}n - 6 = 7 + b$

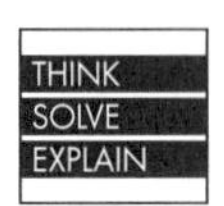


38. Jennifer's age $a$ in 20 years will be less than 39.

 Part A. Write an inequality in simplest form that models the above statement.

 ________________________________

 ________________________________

 ________________________________

 Part B. Label the number line, solve for Jennifer's current age $a$, and graph your inequality.

39. Kayla told her parents she would be back from the senior trip in 14 hours. She became ill and returned in 7 hours 15 minutes. How early was she?

 F. 5 hr and 45 min
 G. 6 hr
 H. 6 hr and 45 min
 I. 7 hr and 45 min

40. A discount store runs a promotion to sell digital cameras at a discounted price. The store uses the equation $0.45b + 20$, where $b$ is the original price, to calculate the discounted price to the customer. A discounted camera sells for \$102. What was the original price to the nearest dollar?

41. Ms. Glover's AP calculus class has 15 students. The students' final averages are displayed on the stem-and-leaf plot. If Irena is in this class, what is the probability that she has a B?

| Grading Scale | |
|---|---|
| 90–100 | A |
| 80–89 | B |
| 70–79 | C |
| 60–69 | D |
| 0–59 | F |

Average Grade

| Stem | Leaf |
|---|---|
| 9 | 0 3 5 7 8 |
| 8 | 3 5 5 7 9 |
| 7 | 5 8 |
| 6 | 0 4 |
| 5 | 8 |

A. $\frac{8}{15}$

B. $\frac{2}{3}$

C. $\frac{1}{3}$

D. $\frac{2}{5}$

42. The histogram shows the hourly wages paid to 14 employees at the Pig-in-a-Poke Restaurant. How many employees earn from 10 to 15 dollars per hour?

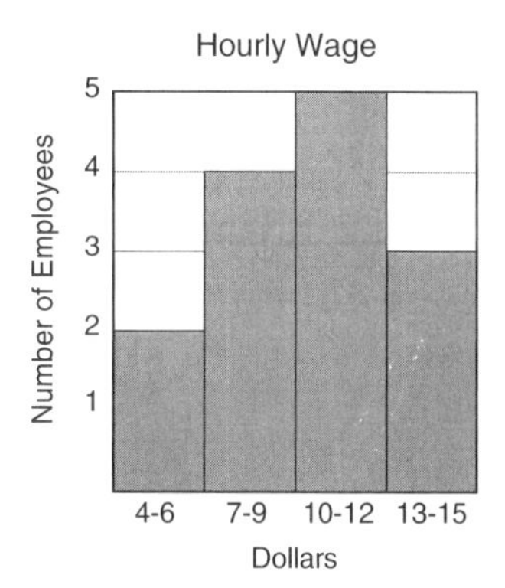

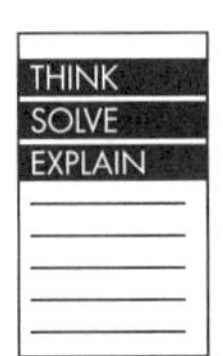


43. Lee Acres Middle School and Anchor Middle School each have a music department with 120 students. The circle graphs show the percentage of students in each school who play a particular instrument.

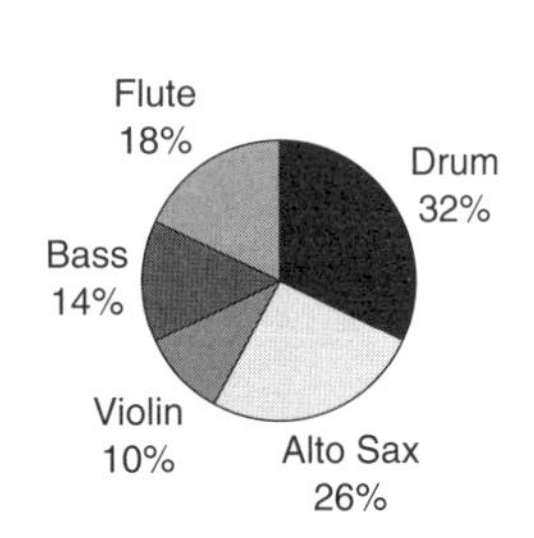


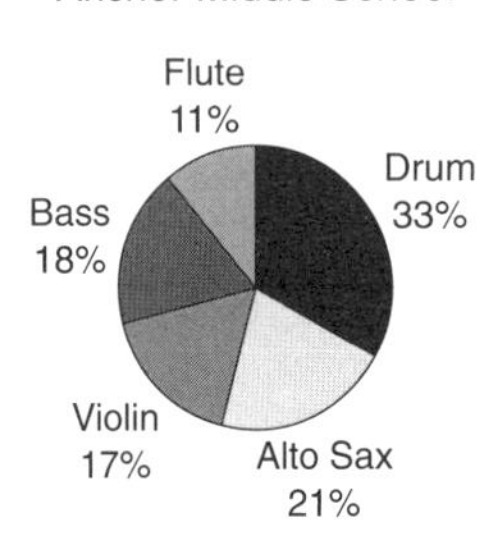


Part A. Use the information from the circle graphs to fill in the tables.

**Lee Acres Middle School**

| Instrument | # of students |
|---|---|
| Flute | |
| Drum | |
| Bass | |
| Violin | |
| Alto Sax | |

**Anchor Middle School**

| Instrument | # of students |
|---|---|
| Flute | |
| Drum | |
| Bass | |
| Violin | |
| Alto Sax | |

Part B. Construct a double-bar graph to display the data for both schools.

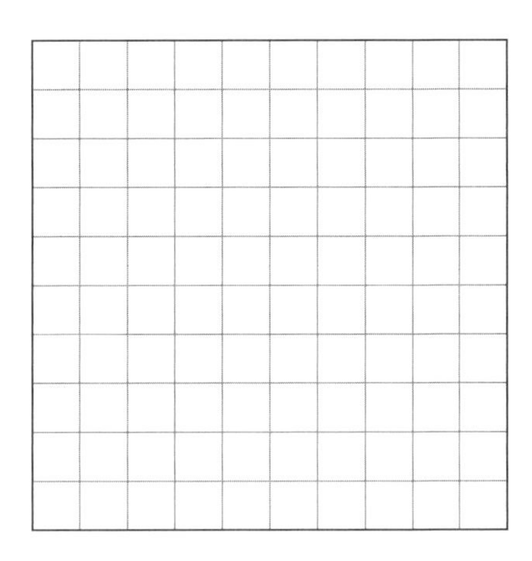

44. Find the range, mean, median, and mode for this data set.

2 cm, 3 cm, 3 cm, 1 cm, 4 cm, 1 cm, 4 cm, 3 cm, 3 cm, and 2 cm.

What number represents three of these measures?

**45.** The mean of Tammie's last five quiz scores was 83. Her median score was 93. What conclusion can you draw from these statistics?

F. She had one very low quiz score.
G. She had one very high quiz score.
H. Her lowest score was 85.
I. Her highest score was 100.

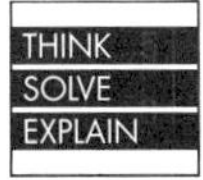


**46.** A traffic light at Toledo Blade in Port Charlotte is programmed according to this pattern: green for 5 minutes and red for 2 minutes. Stuart goes through this intersection twice every day. He wondered about the probability of getting a green light as he drove into the intersection. For 5 days he kept track of the number of times he got a green light ($G$) and the times he got a red light ($R$). Here are the results:
$G\ G\ G\ R\ G\ R\ G\ G\ G\ G.$

Part A. According to the programming, what is the probability that Stuart will get a green light?

________________________________

________________________________

________________________________

Part B. Does the number of times Stuart got a green light match what you would expect from the way the light is programmed? Explain your answer.

________________________________

________________________________

________________________________

**47.** Cindy drives a race car at Pelican Pete's. The probability that she finishes the track ahead of Sam is .44. What are the **odds in favor** of her finishing ahead of Sam?

A. 1 to 44
B. 44 to 1
C. 11 to 14
D. 14 to 11

**48.** The Weather Channel announced that there is a 50% chance of rain today. Which is the best interpretation of this information?

F. It will rain for 12 hours.
G. If it rains, it will rain only at night.
H. It will rain for 50% of the week.
I. The chances that it will rain and that it will not rain today are equal.

49. Jack has invented a new potato peeler. He wants to find out whether people will buy it. Which is the **best** course he could take?

    A. Mail one potato peeler to each of 10,000 people.
    B. Start selling the potato peeler on a small scale, and survey his customers.
    C. Ask people who purchase other kitchen products if they would like to buy his product.
    D. Demonstrate the potato peeler on national television, and survey people who call in.

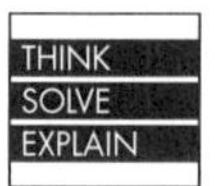


50. The results of a television poll of the watchers of a new program are as shown. An advertising executive concludes that advertising should be directed toward age groups from 15 to 60. Is this sound advice? Explain, using statistical terminology.

15 35 43 45 60

_______________________________________________

_______________________________________________

_______________________________________________

# ANSWERS TO PRACTICE TEST TWO

1. **B.** $\sqrt{730} = 27.02$. D is not correct because 27% in decimal form is 0.27.

2. $\frac{4}{5}$. Since 20 squares out of 25 are shaded, the fraction is $\frac{20}{25}$ or $\frac{4}{5}$.

3. **H.** The number 70,000,000,000 represents 70 billion.

4. **D.** Subtract the exponents when dividing two numbers with the same base: $\frac{10^9}{10^5} = 10^4$.

5. **135.** Multiply 4.5 by 3, which equals 13.5 Then multiply $10^{-7} \times 10^8$, which equals $10^1$ or 10. When 13.5 is multiplied by 10, the result is 135.

6. **I.** Two numbers are additive inverses when their sum equals zero.

7. **A.** Perform all multiplications before subtracting: $87 - 3(12)(2) = 87 - 72$.

8. 4. $\frac{7 \times 2(24 \div 3)}{20 + 14 - 6} = \frac{7 \times 2(8)}{34 - 6} = \frac{7 \times 16}{28} = \frac{112}{28} = 4$

9. **G.** $14\frac{7}{8} - 9\frac{1}{2} = 14.875 - 9.5 = 5.375$ ft, which equals 5 ft $\frac{3}{8}$ in.

10. **2375.** Since 10,000 – 5% = 9500 can be sold in 4 weeks, divide by 4 to find the number than can be sold in 1 week. $9500 \div 4 = 2375$.

11. **370.** There are approximately 370 sq mi for each endangered species. Round 68 to 70. Divide 26,000 by 70.

12. **0.2.** Find the volume of the cube by multiplying: $4 \times 4 \times 4 = 64$ cu in. Multiply 64 by 0.05 to get 3.2 oz for the weight of the tile. Convert 3.2 oz to pounds by dividing by 16: $3.2 \div 16 = 0.2$.

13. Part A. About **25**. Find the volume of the box: $11 \times 8.5 \times 5.5 = 514.25$ cu in. Divide by the number of cubic inches per tangerine: $514.25 \div 20 = 25.71$ or approximately 25.

    Part B. **No.** One cubic foot is equivalent to 12 in × 12 in × 12 in = 1728 cu in. A Priority Mail box holds only 514.25 cu in. Therefore, if Jay fills up his picking basket, a Priority Mail box will not hold all the tangerines.

14. **4050.** There are 45 min in $\frac{3}{4}$ hr. Multiply: $90 \times 45 = 4050$.

15. Part A. Approximately **0.58**. Use a proportion: $\frac{70 \text{ mi}}{60 \text{ min}} = \frac{x \text{ mi}}{0.5 \text{ min}}$ $= 0.5(70)/60$ or about 0.58 mi.

Part B. About **3170**. Round 0.59 to 0.6 mi. Convert to feet by multiplying by 5280, the number of feet in 1 mile: $5280 \times 0.6 = 3168$ ft.

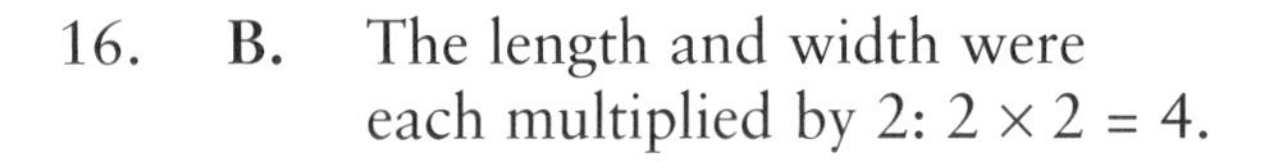

16. **B.** The length and width were each multiplied by 2: $2 \times 2 = 4$.

17. **132.** The original perimeter was $2(10) + 2(12) = 20 + 24 = 44$ ft. The new perimeter will be 3 times the old one: $44 \times 3 = 132$ ft.

18. **H.** Find the scale by dividing the actual measurements by the model measurements:
$\frac{180}{15} = 12$ and $\frac{60}{15} = 12$.

19. **90.** Change fractional measures to decimals and set up a proportion:
$\frac{0.25 \text{ in}}{5 \text{ mi}} = \frac{4.5 \text{ in}}{x}$.
Solve by multiplying $5 \times 4.5$ (= 22.5) and dividing by 0.25 (= 90 mi).

20. **C.** There are 1000 mm in 1 m. Change 4.8 m to millimeters by multiplying by 1000.

21. **1800.** Multiply: $6 \times 5 \times 60 = 1800$ min.

22. **8.** The distance shown in the drawing is a diameter, which is twice the radius.

23. **F.**

24. **A.** Suppose, for example, that quadrilateral *LMPQ* is a trapezoid labeled as shown. Then the only possible answer is *A*.

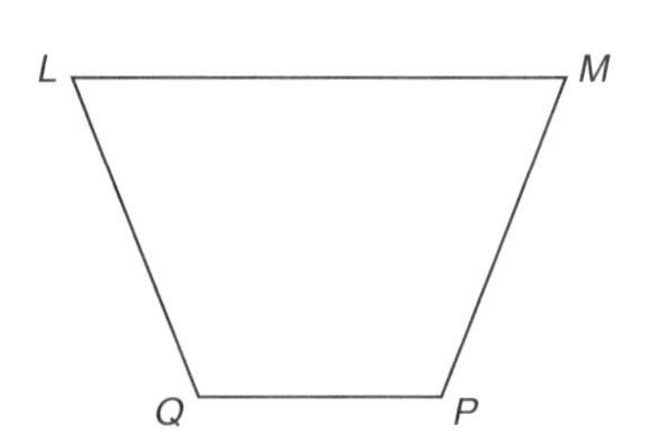


25. **32.** Because the triangles are similar, they are proportional:
$\frac{14}{16} = \frac{28}{x}$.
Multiply 16 by 28, and divide by 14.

26. Part A.

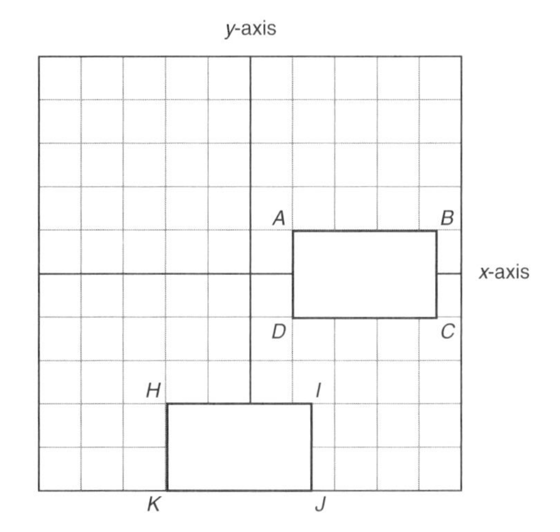


Part B. Place point *A* at (1, 1). Because rectangle *ABCD*, like rectangle *HIJK*, is 3.5 spaces long, it must go to the right so that it won't be in Quadrant II. It is 2 spaces wide; therefore, if it's drawn to look like *HIJK*, the *x*-axis will cut it into two exactly similar parts.

27. **F.**

28. **A.** A line of symmetry cuts a figure into two equal parts that, when folded together over the line, will match exactly.

29. Part A. Lightning *A* translates (slides) right 2 spaces and down 3 spaces to become lightning *B*. Lightning *B* translates (slides) right 4 spaces and down 3 spaces to become lightning *C*.

Part B. **5.** The distance from lightning *B* to lightning C is 3 vertical spaces and 4 horizontal spaces. The distance between *B* and C is found using the Pythagorean theorem: $a^2 + b^2 = c^2$. Substitute $a = 3$ and $b = 4$ to find *c*, the distance.

$a^2 + b^2 = c^2$
$3^2 + 4^2 = c^2$
$9 + 16 = c^2$
$25 = c^2$
$\sqrt{25} = c = 5$

30. **G.** This line has one *x*-intercept and has undefined slope.

$\text{Slope} = \dfrac{\text{change in } y}{\text{change in } x}.$

Here, the change in *x* is zero. Division by zero is undefined; therefore the slope is undefined.

31. Part A.

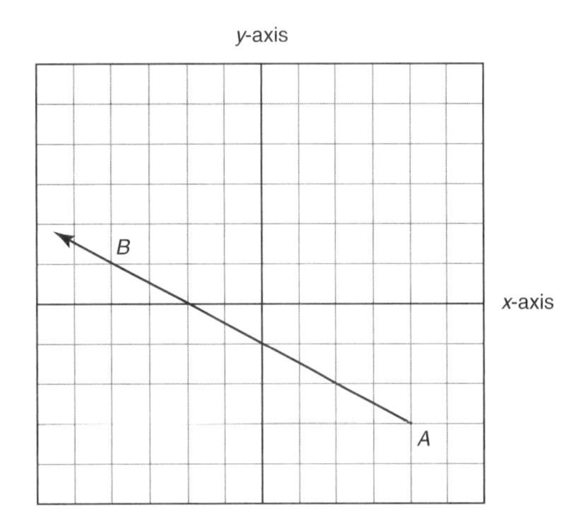


Part B. Quadrant **II** contains *B*, and Quadrant **IV** contains *A*.

32. **C.** Substitute the *x* and *y* values in the given inequality:

$y \le 4x - 6$
$12 \le 4(4) - 6$
$12 \le 16 - 6$
$12 \le 10$ 12 is not less than or equal to 10.

33. **7.** It helps to make a table of the values:

| Triangles | Squares |
|---|---|
| 1 | 2 |
| 2 | 5 |
| 3 | 8 |
| 4 | 11 |
| 5 | 14 |
| 6 | 17 |
| 7 | 20 |

34. **G.** The line has a $y$-intercept of –2 and a slope of $-\frac{1}{2}$.

35. **1.5.** Substitute the given value of $x$ into the equation:
$y = 15(0.5) - 6 = 7.5 - 6 = 1.5$.

36. Part A. **–5000.**
Part B. A slope of 5000 means that the temperature drops by 1° for every 5000 ft of depth. The slope of the equation of the water rise (depth) over the run (temperature) is $\frac{4000-5000}{3.8-3.6} = \frac{-1000}{.2} = -5000$.

37. **B.**

38. Part A. $\boldsymbol{a + 20 < 39}$.

Part B. $a < 39 - 20$.
$\boldsymbol{a < 19}$.

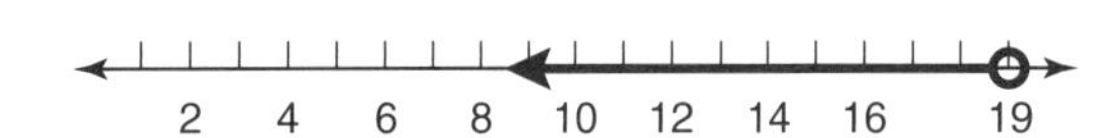


39. **H.** Subtract: 14 – 7 = 7 hr. Kayla used an additional 15 min from the 7 hr.

40. **182.** Find the original price by setting up the equation:
$102 = 0.45b + 20$.
Subtract 20 from both sides:
$102 - 20 = 0.45b + 20 - 20$.
Simplify: $82 = 0.45b$.
Divide by 0.45: $82 \div 0.45 = 182.22$.

41. **C.** Five students out of 15 received grades of B: $\frac{5}{15} = \frac{1}{3}$.

42. **8.** Five employees receive \$10–12, and 3 receive \$13–15, for a total of 8 employees receiving \$10–15 per hour.

43. Part A. Fill in the tables by multiplying 120 by each percent. For example, to find the number of students at Lee Acres Middle School who play the flute, multiply $120 \times 18\% = 21.6$. Round to the nearest whole number: 21.6 = 22.

**Lee Acres Middle School**

| Instrument | # of students |
|---|---|
| Flute | 22 |
| Drum | 38 |
| Bass | 17 |
| Violin | 12 |
| Alto Sax | 31 |

**Anchor Middle School**

| Instrument | # of students |
|---|---|
| Flute | 13 |
| Drum | 40 |
| Bass | 22 |
| Violin | 20 |
| Alto Sax | 25 |

Part B.

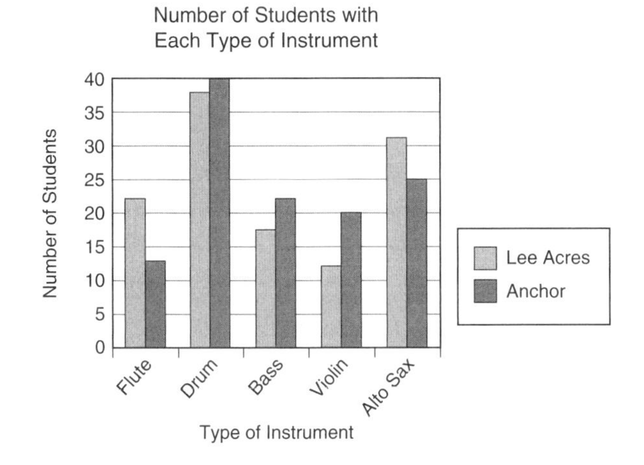


Part B. **No.** Stuart's results show 8 green lights out of a total of 10 possible lights: $\frac{8}{10}$. If you express $\frac{5}{7}$ and $\frac{8}{10}$ as decimals, you can compare them more easily: $\frac{5}{7} = .71$ and $\frac{8}{10} = .8$. These values are fairly close, but for a nearly perfect match Stuart should have had 3 red lights out of ten (or 7 green lights out of 10).

44. **3.** Range = **3**, mean = **2.6**, median = **3**, mode = **3**. **3** represents range, median, and mode.

45. **F.** If Tammie's median score was 93, then she had two quiz scores above 93. In order for the mean to be 83, she must have had one very low quiz score to drop the average. Not enough information is given to assume that she had 100 on any test.

46. Part A. $\frac{5}{7}$ For a total of 7 min, the probability of a green light is $\frac{5}{7}$.

47. **C.** If the probability in favor is 0.44, the probability against is 0.56. Expressed as a ratio, 44:56 = 11:14 in lowest terms.

48. **I.**

49. **D.** Demonstrating on national television gives more people the opportunity to see and evaluate the product.

50. **No.** The advertising executive misunderstands the range of the data and ignores the quartiles, which indicate that half the viewers are between the ages of 35 and 45.

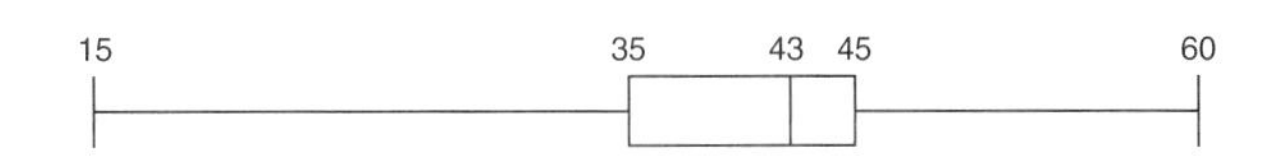

# Index